TIME AND THE EARTH'S ROTATION

INTERNATIONAL ASTRONOMICAL UNION
UNION ASTRONOMIQUE INTERNATIONALE

SYMPOSIUM No. 82

PROCEEDINGS OF THE 82ND SYMPOSIUM OF THE INTERNATIONAL
ASTRONOMICAL UNION HELD IN SAN FERNANDO, SPAIN, 8-12 MAY, 1978

TIME AND THE EARTH'S ROTATION

EDITED BY

DENNIS D. McCARTHY

U.S. Naval Observatory, 34th and Massachusetts Avenue N.W., Washington, D.C., U.S.A.

and

JOHN D. H. PILKINGTON

Royal Greenwich Observatory, Herstmonceux Castle, Hailsham, East Sussex, England

D. REIDEL PUBLISHING COMPANY

DORDRECHT : HOLLAND / BOSTON : U.S.A. / LONDON : ENGLAND

Library of Congress Cataloging in Publication Data

Main entry under title:

Time and the Earth's rotation.

(Symposium – International Astronomical Union; no. 82)
Papers presented at a symposium held in San Fernando, Spain, May 8–12, 1978.
Includes index.
1. Time–Congresses.   2. Earth–Rotation–Congresses.   I. McCarthy, Dennis D.
II. Pilkington, J. D.   III. Series: International Astronomical Union. Symposium; no. 82.
QB209.T523   525'.35   79–14967
ISBN 90-277-0892-4
ISBN 90-277-0893-2 pbk.

Published on behalf of
the International Astronomical Union
by
D. Reidel Publishing Company, P.O. Box 17, Dordrecht, Holland

All Rights Reserved
Copyright © 1979 by the International Astronomical Union

Sold and distributed in the U.S.A., Canada, and Mexico
by D. Reidel Publishing Company, Inc.
Lincoln Building, 160 Old Derby Street, Hingham,
Mass. 02043, U.S.A.

No part of the material protected by this copyright notice may be reproduced or utilized in any form or by any means, electronic or mechanical, including photocopying, recording or by any informational storage and retrieval system, without written permission from the publisher

Printed in The Netherlands

TABLE OF CONTENTS

Preface   xi

Organizing Committees   xii

Acknowledgements   xiii

List of Participants   xv

INTRODUCTION: R. O. VICENTE / The Future of the Researches   1
    About the Earth's Rotation

I. TIME

B. GUINOT / Basic Problems in the Kinematics of the Rotation
    of the Earth (Invited Paper)   7

S. PUSHKIN / A New State Time and Frequency Standard of the
    USSR   19

D. YU. BELOTSERKOVSKIJ and M. B. KAUFMAN / A New Method of
    Universal Time Computation Used in the State Time
    and Frequency Service of the USSR   23

YA. S. YATSKIV, A. A. KORSUN', and N. T. MIRONOV / On the
    Determination of UT1 by the BIH and the U.S.S.R. Time
    Service   29

S. DEBARBAT / Etude d'Observations Effectuées à l'Astrolabe de
    Paris et Comparaison avec d'Autres Resultats Concernant
    les Termes Principaux de la Nutation   41

V. S. GUBANOV and L. I. YAGUDIN / A New System of the U.S.S.R.
    Standard Time for 1955-1974 and its Application in the
    Study of the Earth's Rotation   47

D. D. MCCARTHY and D. B. PERCIVAL / An Analysis of the
    Rotational Acceleration of the Earth (Abstract)   53

L. V. MORRISON / New Determination of the "Decade" Fluctuations
    in the Rotation of the Earth, 1860-1978 (Abstract) 55

A. I. EMETZ and A. A. KORSUN' / On the Long-Period Variations
    in the Rate of the Earth's Rotation (Abstract)   59

N. S. SIDORENKOV / Main Results of Studying the Nature of the
    Irregularity of the Earth's Rotation   61

D. D. MCCARTHY / Some Advantages and Disadvantages of a
  Photographic Zenith Tube (abstract) 65

N. P. J. O'HORA and T. F. BAKER / Tidal Perturbations in
  Astronomical Observations 67

J. POPELAR / Polar Motion and Earth Rotation Monitoring
  in Canada 73

D. DJUROVIC / Polar Coordinates and UT1-UTC from PZT
  Observations 75

S. IIJIMA, S. FUJII, and Y. NIIMI / (α-2L) Terms as
  Obtained from PZT Observations 79

G. BILLAUD / Etude Succinte du Catalogue Fondamental FK4
  à Partir des Observations Faites à l'Astrolabe 85

II. POLAR MOTION

E. P. FEDOROV / On the Coordinate Systems Used in the Study
  of Polar Motion (Invited Paper) 89

S. YUMI, K. YOKOYAMA, and H. ISHII / Derivation of Pole
  Coordinates in a Uniform System from the Past ILS Data 103

M. FEISSEL / On the Computation of Accurate Earth Rotation
  by the Classical Astronomical Method 109

A. POMA and E. PROVERBIO / On the Relative Motion of the
  Earth's Axis of Figure and the Pole of Rotation 115

V. P. SHCHEGLOV and G. M. KAGANOVSKY / Secular Variation
  of Tashkent Astronomical Latitude (Abstract) 123

B. KOLACZEK / Coordinates of the Pole for the Period
  1968-1974 Computed in the System of 10 Stations with
  Small Variations of Mean Latitudes 125

F. CHOLLET / Amelioration des Calculs de Reduction des
  Observations à l' Astrolabe. Application à la
  Determination des Termes de 18.6 et 9.3 Ans de la
  Nutation 129

L. M. BARRETO / Time and Latitude Programs at the National
  Observatory of Brazil 135

L. BUFFONI, F. CARTA, F. CHLISTOVSKY, A. MANARA, and
F. MAZZOLENI / Preliminary Analysis of Astrolabe Observations
  at Merate Observatory During the Period 1970-1977
  (Abstract) 137

TABLE OF CONTENTS

K. KANIUTH and W. WENDE / The Longitude Difference Merate-Milano Derived from Danjon Astrolabe Observations by Means of a One-Step Adjustment Using an Extended Model — 139

J. MOCZKO / Velocity of the Motion of the Terrestrial Pole — 145

III. REFERENCE SYSTEMS

J. KOVALEVSKY / The Reference Systems (Invited Paper) — 151

C. A. MURRAY / The Ephemeris Reference Frame for Astrometry — 165

N. CAPITAINE / Nutation in Space and Diurnal Nutation in the Case of an Elastic Earth — 169

E. W. GRAFAREND, I. I. MUELLER, H. B. PAPO, and B. RITCHER / Concept for Reference Frames in Geodesy and Geodynamics: The Reference Directions (Abstract) — 175

IV. RADIO INTERFEROMETRY

B. ELSMORE / An Introduction to Radio Interferometric Techniques — 177

K. J. JOHNSTON / The Application of Radio Interferometric Techniques to the Determination of Earth Rotation (Invited Paper) — 183

W. E. CARTER, D. S. ROBERTSON, and M. D. ABELL / An Improved Polar Motion and Earth Rotation Monitoring Service Using Radio Interferometry — 191

J. L. FANSELOW, J. B. THOMAS, E. J. COHEN, P. F. MACDORAN, W. G. MELBOURNE, B. D. MULHALL, G. H. PURCELL, D. H. ROGSTAD, L. J. SKJERVE, and D. J. SPITZMESSER / Determination of UT1 and Polar Motion by the Deep Space Network Using Very Long Baseline Interferometry — 199

K. J. JOHNSTON, J. H. SPENCER, C. H. MAYER, W. J. KLEPCZYNSKI, G. KAPLAN, D. D. MCCARTHY, and G. WESTERHOUT / The NAVOBSY/NRL Program for the Determination of Earth Rotation and Polar Motion — 211

D. S. ROBERTSON, W. E. CARTER, B. E. COREY, W. D. COTTON,
C. C. COUNSELMAN, I. I. SHAPIRO, J. J. WITTELS,
H. F. HINTEREGGER. C. A. KNIGHT, A. E. E. ROGERS,
A. R. WHITNEY, J. W. RYAN, T. A. CLARK, R. J. COATES,
C. MA, and J. M. MORAN / Recent Results of Radio Interferometric Determinations of a Transcontinental Baseline, Polar Motion, and Earth Rotation 217

H. G. WALTER / Precision Estimates of Universal Time from Radio-Interferometric Observations 225

V. SATELLITE LASER RANGING

D. E. SMITH, R. KOLENKIEWICZ, P. J. DUNN, and M. TORRENCE / Determination of Polar Motion and Earth Rotation from Laser Tracking of Satellites (Invited Paper) 231

B. E. SCHUTZ, B. D. TAPLEY, and J. RIES / Polar Motion from Laser Range Measurements of GEOS-3 239

P. L. BENDER and E. C. SILVERBERG / Low Cost Lageos Ranging System (Abstract) 245

VI. LUNAR LASER RANGING

E. C. SILVERBERG / On the Effective Use of Lunar Ranging for the Determination of the Earth's Rotation (Invited Paper) 247

J. D. MULHOLLAND / Is Lunar Ranging a Viable Component in a Next-Generation Earth Rotation Service? 257

O. CALAME / Preliminary UTO Results from EROLD Data 261

VII. DOPPLER SATELLITE METHODS

C. OESTERWINTER / Polar Motion through 1977 from Doppler Satellite Observations (Invited Paper) 263

B. GUINOT / Irregularities of the Polar Motion 279

P. PAQUET and C. DEVIS / Reasons and Possibilities for an Extended Use of the Transit System 287

F. NOUEL and D. GAMBIS / Very First Results of the MEDOC Experiment 295

VIII. GEOPHYSICS

S. K. RUNCORN / The Geophysical Interpretation of Changes in the Length of the Day and Polar Motion 301

C. R. WILSON / Estimation of the Parameters of the Earth's Polar Motion 307

S. TAKAGI / Rotational Velocity of an Earth Model with a Liquid Core (Abstract) 313

G. P. PIL'NIK / On Investigations of the Tidal Waves M_f and M_m (Abstract) 315

P. BROSCHE and J. SUNDERMANN / Oceanic Tidal Friction: Principles and New Results 317

V. P. SHCHEGLOV / The Stability of Continental Blocks in Seismically Active Regions (Abstract) 321

GENERAL DISCUSSION 323

RESOLUTIONS 329

INDEX **331**

PREFACE

IAU Symposium No. 82, "Time and the Earth's Rotation", met to discuss modern research in the field of the rotation of the Earth with particular emphasis on the role of new observational techniques in this work. The use of these techniques has prompted a new look at the definitions of the traditional reference systems and the concepts of the rotation of the Earth around its center of mass. Specific topics discussed were time, polar motion, reference systems, conventional radio interferometry, very long baseline interferometry (VLBI), Doppler satellite methods, satellite laser ranging, lunar laser ranging, and geophysical research concerning the Earth's rotation.

Improvement in the accuracy of the observations is a key to possible solutions of the many unsolved problems remaining in this field. It appears that such improvement, using both classical and new techniques, is forthcoming in the near future. This will surely contribute to a better understanding of some of the long-standing questions concerning the rotation of the Earth around its center of mass and lead to an improved knowledge of the rotating, deformable Earth.

This volume contains the papers presented at IAU Symposium No. 82 as well as the discussions provoked by these papers. It is hoped that it captures the principal points of the meeting and that it will contribute not only to a better understanding of existing problems, but also to future research in time and the Earth's rotation.

SCIENTIFIC ORGANIZING COMMITTEE

R. O. Vicente, Chairman

B. Elsmore (UK)

B. Guinot (France)

S. Iijima (Japan)

K. Johnston (USA)

D. D. McCarthy (USA)

A. Orte (Spain)

P. E. Pâquet (Belgium)

J. D. Pilkington (UK)

H. M. Smith (UK)

Y. S. Yatskiv (USSR)

S. Yumi (Japan)

LOCAL ORGANIZING COMMITTEE

A. Orte, Chairman

M. Sanchez, Secretary

J. Benavente, Member

M. Catalan, Member

L. Quijano, Member

ACKNOWLEDGEMENTS

Many people contributed to the success of this symposium. In particular we would like to thank the Spanish National Committee for Astronomy and the Spanish Navy for their support, and the Local Organising Committee and the staff of the Instituto y Observatorio de Marina for conducting the meetings, collecting and assembling the papers and the records of the discussions, and arranging a most enjoyable social programme. The editors gratefully acknowledge the help of Mrs Dorothy Outlaw and Miss Alice Babcock of the U.S. Naval Observatory and Miss Hilary Saunders of the Royal Greenwich Observatory in the preparation of this volume.

LIST OF PARTICIPANTS

BARCIA, A., Observatorio Astronomico, Alfonso XII, 3, Madrid-7, Spain.

BARRETO, L. M., Observatorio Nacional, R. General Bruce 586, 20000 Rio de Janeiro, Brazil.

BENAVENTE, J., Inst. y Observatorio de Marina, San Fernando (Cadiz), Spain.

BILLAUD, G., CERGA, Avenue Copernic, 06130 Grasse, France.

BOLOIX, M., Observatorio de Marina, San Fernando (Cadiz), Spain.

BREGMAN, J. D., Netherlands Fund. Radio Astr., Westerbork, Post Hooghalen, Netherlands.

BROSCHE, P., Observatorium Hoher List, Universitats-Sternwarte Bonn, D-5568 Daun, Germany F.R.

CALAME, O., CERGA/ROQUEVIGNON, Av. Copernic, 06130 Grasse, France.

CAMPBELL, J., Geodatisches Institut, Nussallee 17, 53 Bonn, Germany F.R.

CAPRIOLI, G., Osservatorio Astronomico di Roma, Via Trionfale 204, Rome, Italy.

CARTER, W. E., National Geodetic Survey, C-133, Rockville, MD 20852, USA.

CATALAN, M., Observatorio de Marina, San Fernando (Cadiz), Spain.

CHLISTOVSKY, F., Observatorio Astronomico di Brera, Via Brera, 28, 20121 Milano, Italy.

CHOLLET, F., Observatoire de Paris, 61 Av. de l'Observatoire, 75014 Paris, France.

CONEJERO FERNANDEZ, G. J., Serv. Geographico del Ejercito, Prim, 8, Madrid-4, Spain.

DE ORUS NAVARRO, J. J., Cat. Astronomia Facultad Fisicas, Avda. Gral. Franco 647, Universidad, Barcelona-28, Spain.

DEBARBAT, S., Observatoire de Paris, 61 Av. de l'Observatoire, 75014 Paris, France.

DJUROVIC, D., Belgrade University, Volgina 7, 11050 Beograd, Yugoslavia.

LIST OF PARTICIPANTS

DOW, J. M., European Space Operations Center, Robert-Bosch-Strasse 5, 6100 Darmstadt, Germany F.R.

ELSMORE, B., Cavendish Laboratory, Madingly Road, Cambridge CB3 OHE, England.

ENSLIN, H., Deutsches Hydrographishes Inst., Postfach 220, 2000 Hamburg 4, Germany F.R.

FANSELOW, J. L., Jet Propulsion Laboratory, 4800 Oak Grove Drive, Pasadena, CA 91103, USA.

FEDOROV, E. P., Main Astron. Obs., Ukrainian Academy of Sciences, 252127 Kiev 127, USSR.

FEISSEL, M., Bureau International de l'Heure, 61 Av. de l'Observatoire, 75014 Paris, France.

GAMBIS, D., CRGS/CNES, 18 Rue Edouard Belin, 31055 Toulouse-Cedex, France.

GARCIA DE POLAVIEJA, M., Observatorio de Marina, San Fernando (Cadiz), Spain.

GOMEZ ARMARIO, F., Observatorio de Marina, San Fernando (Cadiz), Spain.

GOMEZ GONZALEZ, J., Observatorio Astronomico, Alfonso XII, 3, Madrid-7, Spain.

GUBANOV, V. S., Pulkovo Observatory, Leningrad 196140, USSR.

GUINOT, B., Bureau International de l'Heure, 61 Av. de l'Observatoire, 75014 Paris, France.

HATAT, J. L., Observatorio de Marina, San Fernando (Cadiz), Spain.

IIJIMA, S., Tokyo Astronomical Observatory, Mitaka, Tokyo, 181, Japan.

JOHNSTON, K., Code 7134, Naval Research Laboratory, Washington, DC 20375, USA.

KANIUTH, K., Deutsches Geodat. Forschungsinst., Marstallplatz 8, 8000 Munchen 22, Germany F.R.

KOLACZEK, B., Planetary Geod. Dept. S.R.C. PAS, Palac Kultury i Nauki 2313, 00-901 Warszawa, Poland.

KOVALEVSKY, J., CERGA, 8 Blvd. Emile Zola, 06130 Grasse, France.

MANARA, A., Observatorio Astronomico di Brera, Via Brera 28, 20121 Milano, Italy.

LIST OF PARTICIPANTS

MCCARTHY, D. D., U. S. Naval Observatory, Washington, DC 20390, USA.

MELBOURNE, W. G., Jet Propulsion Laboratory, 4800 Oak Grove Dr., Bldg 264-747, Pasadena, CA 91103, USA.

MIGUEL LAFUENTE, T., Instituto Geografico Nacional, General Ibanez de Ibero 3, Madrid-3, Spain.

MORRISON, L. V., Royal Greenwich Obs., Herstmonceux Castle, Hailsham, Sussex BN27 1RP, England.

MUELLER, I. I., Dept. of Geodetic Science, Ohio State University, Columbus, OH 43210, USA.

MUINOS, J. L., Observatorio de Marina, San Fernando (Cadiz), Spain.

MULHOLLAND, J. D., Dept. of Astronomy, University of Texas, Austin, TX 78712, USA.

MURRAY, C. A., Royal Greenwich Observatory, Herstmonceux Castle, Hailsham, Sussex, England.

NAKAJIMA, K., Tokyo Astronomical Observatory, Mitaka, Tokyo, 181, Japan.

NOUEL, F., CRGS/CNES, 18 Av. Edouard Belin, 31055 Toulouse Cedex, France.

O'HORA, N.P.J., Royal Greenwich Observatory, Herstmonceux Castle, Hailsham, Sussex, England.

OESTERWINTER, C., Naval Surface Weapons Center, Dahlgren Laboratory, Code CK-101, Dahlgren, VA 22448, USA.

ORTE, A., Observatorio de Marina, San Fernando (Cadiz), Spain.

OTERMA, L., Astron.-Opt. Inst., Univ. Turku, Sirkkalankatu 31, 20700 Turku 70, Finland.

PAQUET, P.E.G., Observatoire Royal de Belique, Av. Circulaire 3, B-1180 Bruxelles, Belgium.

PENSADO IGLESIAS, J., Observatorio Astronomico, Alfonso XII, 3, Madrid-7, Spain.

PILKINGTON, J., Royal Greenwich Observatory, Herstmonceux Castle, Hailsham, Sussex BN27 1RP, England.

PINTO CORDERO, G., Instituto Geografico Nacional, General Ibanez Ibero 3, Madrid-3, Spain.

POMA, A., Stazione Astronomica Latitudne, Via Ospedale 72, 09100 Cagliari, Italy.

PONS ALCANTARA, J. M., Serv. Geografico del Ejercito, Prim 8, Madrid-4, Spain.

POPELAR, J., Earth Physics Branch D.E.M.R., 3 Obs. Crescent, Ottawa, Ontario, KIA OE4, Canada.

PROVERBIO, E., Institute of Astronomy, Via Ospedale 72, 09100 Cagliari, Italy.

QUIJANO, L., Inst. Observatorio de Marina, San Fernando (Cadiz), Spain.

RIUS JORDAN, A., INTA/NASA, Orense 11, Madrid-20, Spain.

ROBBINS, A. R., Dep. of Surveying and Geodesy, University of Oxford, 62 Banbury Road, Oxford OX2 6PN, England.

RUNCORN, S. K., School of Physics, University of Newcastle, Newcastle Upon Tyne NE1 7RU, England.

SALAZAR, A., Observatorio de Marina, San Fernando (Cadiz), Spain.

SANCHEZ, M., Observatorio de Marina, San Fernando (Cadiz), Spain.

SCHLUTER, W., Institut Angewandte Geodasie, Weinbergstr. 9, D-6230 Frankfurt-Sindlingen, Germany F.R.

SCHUTZ, B. E., University of Texas, Dept. of Aerospace Engineering, Austin, TX 78712, USA.

SHCHEGLOV, V. P., Astronomical Observatory, Tashkent, USSR.

SILVERBERG, E. C., Dept. of Astronomy, University of Texas, Austin, TX 78712, USA.

SMITH, D. E., Code 921, Goddard Space Flight Center, Greenbelt, MD 20771, USA.

SMITH, H. M., 23 Normandale, Bexhill on Sea, East Sussex, TN39 3LU, England.

TAKAGI, S., 59-1, Shinjo-Hamaba, Mizusawa-Shi, Iwate-Ken, 023 Japan.

TAPLEY, B. D., Dept. Aerospace Engineering, University of Texas, Austin, TX 78712, USA.

TORROJA MENENDEZ, J. M., Universidad Complutense, Ciudad Universitaria, Madrid-3, Spain.

LIST OF PARTICIPANTS

TSUCHIYA, A., Tokyo Astron. Obs., University of Tokyo, Mitaka, Tokyo, 181, Japan.

TUR SERRA, G. R., Serv. Geografico del Ejercito, Prim 8, Madrid-4, Spain.

VALEIN, J. L., CERGA/ROQUEVIGNON, Av. Nicolai Copernic, 06130 Grasse, France.

VICENTE, R. O., R. Mestre Aviz, 30, R/C, Lisboa 3, Portugal.

VIEIRA DIAZ, R., Fac. Ciencias Matematicas Astr., Universidad Complutense, Madrid-3, Spain.

WALTER, H. G., Astronomisches Rechen-Institut, Monchhofstr. 12-14, D-6900 Heidelberg 1, Germany F.R.

WENDE, W., Deutsches Geodat. Forschungsints., Marstallplatz 8, 8000 Munchen 22, Germany F.R.

WIETH-KNUDSEN, N. P., Svend Trostsvej 12 IV, 1912 Copenhagen-V, Denmark.

WILKINS, G. A., Royal Greenwich Observatory, Herstmonceux Castle, Hailsham, Sussex BN27 1RP, England.

WILSON, C. R., Dept. of Geological Sciences, University of Texas, Austin, TX 78712, USA.

YATSKIV, Y. S., Main Astron. Obs., Ukrainian Academy of Sciences, 252127 Kiev 127, USSR.

YOKOYAMA, K., Int. Latitude Obs. of Mizusawa, Mizusawa-Shi, Iwate-Ken, 023 Japan.

YUMI, S., Int. Latitude Obs. of Mizusawa, Mizusawa-Shi, Iwate-Ken, 023 Japan.

INTRODUCTION

THE FUTURE OF THE RESEARCHES ABOUT THE EARTH'S ROTATION

R. O. Vicente
Department of Applied Mathematics, Faculty of Sciences,
Lisbon University, Portugal.

It is well known that futurology, the science of forecasting the future, is a very difficult science. It also depends on the intervals of time for which we pretend to forecast the evolution of the physical system under consideration. Aldous Huxley thought that futurology might be a very good thing.

We are concerned about the future of the researches dealing with the Earth's rotation and I shall try a few guesses, considering suitable time intervals.

The astronomical and geological evolution of our planet can be conveniently expressed in units of a million years, but this time interval is very long from the point of view of the researches of the human civilisations dealing with the dynamics of the Earth around its centre of mass. Even a thousand years is already too long to attempt any forecast of our researches, because we cannot guess if our present type of civilisation will last as long as that; but we know that human civilisations have observed the luni-solar precession for about 2,000 years. If we consider one hundred years we are already in a better position to guess the future of our researches, but we must not forget that futurology is a very hazardous science. Actually the first systematic observations of the phenomena that concern us were made about a hundred years ago; that is, we are at the end of the first century of observations dealing with the Earth's rotation.

Another way to look into the future of our researches, for the next hundred years, is to study the history of science dealing with this subject during the last hundred years. The consideration of the main fields of advance in the researches concerned with the latitude and longitude variations depends, sometimes, on the viewpoint of the person who is looking into the subject.

About a hundred years ago, a large section of the astronomical community was sceptical about the possibility of measuring the small variations of latitude that were predicted when the dynamical theory of the motion of a rigid body around its centre of mass was applied to the case of the Earth. But patient and determined observers, employing the best instruments available at the time, suceeded in determining some variations of latitude.

This pioneer work was followed by the remarkable studies of Chandler that finally convinced everybody of the possibility of determining latitude variations. The so-called Chandler period corresponds to the period of the free nutation of the Earth. This observed value refers to the real Earth, and the theoretical values computed from mathematical models should therefore be in agreement with these observations. At that time the knowledge about the internal structure of our planet was very scanty and geophysics was beginning to appear as a science.

The successful demonstration of the possibility of determining variations of latitude led to the setting up of a remarkable programme of astronomical international cooperation called the International Latitude Service. The first observations were made in December 1899 and have continued, without interruptions, till the present day.

Another aspect of the motion of the Earth around its centre of mass corresponds to the study of the diurnal rotation and its irregularities; that is, the time problems. The theory we have mentioned forecasts the constancy of the 24 hours period to such a degree of precision that observation of irregularities of the daily rotation was not technically feasible until much later.

While latitude variations were already regularly observed at the beginning of the century, thanks to the development of adequate techniques, the corresponding time variations had to wait for further development of timekeeping devices. We must remember that pendulum clocks were still the main timekeeping instruments till about 40 years later. The researches of N. Stoyko, based on pendulum clocks, first detected the irregularities in the length of the day.

The organisation of another important programme of international cooperation, called the Bureau International de l'Heure, has been of fundamental importance not only in the adoption and improvement of new timekeeping devices but also in theoretical researches dealing with time problems.

These facts show the great importance that technological developments have on further improvements of our researches.

As we try to guess the future by looking into the past, let us define which were the main developments. The most important one, I think, was the establishment of international cooperation on a regular and permanent basis. Another important milestone was the adoption of the same type of instrument, and the same computing technique, for the reduction of latitude observations. When we remember that logarithms

INTRODUCTION

were the main computing technique and we see, nowadays, our electronic computers, we can guess that our colleagues, in a hundred years' time, will probably consider our computing techniques as primitive as we consider logarithms at the present time.

The theory employed in those days to forecast latitude and time variations can be considered as simple in comparison with the present day standards. The main difficulties were lack of adequate knowledge about the structure of the Earth, and lack of computing techniques that facilitated the integration of the equations of motion describing the dynamics of the Earth.

The advances in internal and external geophysics, specially during the last 50 years, gave us a good insight about the Earth. The progress of seismology and the publication of the seismological tables of Jeffreys and Bullen, based on proper statistical analysis, permitted the definition of the main layers of the Earth's interior. The knowledge about our atmosphere and hydrosphere also contributes to a better understanding of polar motion and time irregularities.

What will happen in the next century? It is known that some branches of science, after rapid strides in a fairly short time interval, tend to slow down for a time till new technological and theoretical advances are possible. So the answer to this question depends on the optimistic or pessimistic outlook of the person who tries to guess the future. My guess is that internal geophysics will not advance in a way that will modify substantially the important influence it has on time and polar motion problems, but external geophysics might show appreciable improvements in the next hundred years.

The present-day computing techniques have allowed us to set up theoretical models that take account of the actual knowledge about the structure of the Earth. But, precisely because of the easy access to computers, there is such a proliferation of theoretical models that it is extremely difficult to compare results; this applies not only for the structure but also for the geopotential, which is so important for the computation of orbits of the Earth satellites.

I hope that this problem will be solved in the near future by the setting up of reference models for the different regions, from the centre to the outer layers of the atmosphere, that make up our planet. The application of Rayleigh's principle, to calculate generally adopted first order perturbations of the reference models, will increase the usefulness of such models.

We are forgetting one of the main lessons of the past that I have already mentioned, namely, the need for adopting the same computing procedures. This is still more important nowadays than at the beginning of the century, because our computer programmes are so complex that it is not easy for an outsider to check the computations; that was an advantage of our colleagues of a hundred years ago, thanks to their primitive computing techniques. I hope that our colleagues in a hundred years' time will have adopted more streamlined and sophisticated computing techniques that will allow anybody to check their computations more easily.

I have already mentioned that latitude variations were observed far earlier than time variations; but technological advances, made around the middle of the present century, led to the construction of quartz and atomic clocks of such precision that the daily rotation of the Earth, our fundamental clock since the first human beings appeared on our planet, is no longer employed in many branches of physics. As, however, most of us are bound to the surface of this planet, we are still obliged to compare these timekeeping devices - the atomic clock and the diurnal motion of the Earth. There are a number of irregularities in time observations that we are beginning to study, thanks to the high precision attained by the atomic clocks, and these observations will oblige theoreticians to set up models trying to explain the irregularities.

What will the future reserve for us in this field of research? We had a rapid technological development of timekeeping devices for about 20 years, and what will happen in the next hundred years? I leave the questions without answer, but I think the explanation of these small irregularities will keep scientists busy for quite a time.

We are at present experiencing another technological revolution in the field of studies related to the motion of the Earth around its centre of mass. This is based on the utilization of Doppler and laser techniques with artificial satellites, laser techniques with the Moon, and very long base line interferometry with radio telescopes. These are the so-called modern techniques, which should be called the present-day techniques in the context of the viewpoint we are considering because they will probably be superseded in a hundred years.

We are debating, at the moment, the merits of these modern techniques in comparison with the classical techniques employing visual zenith telescopes (VZT), photographic zenith telescopes (PZT) and astrolabes. There are also discussions about the advantages and disadvantages of the above mentioned techniques for polar motion and universal time determinations.

We are at an important turning point in the researches about the Earth's rotation, thanks to this technological revolution. We can see that the classical techniques will be superseded, in due time, by modern techniques but one important matter, under discussion at present, is to know the right moment when the classical techniques might be discontinued, or if some of these techniques (for instance, the PZT) will still be useful in the next decades.

INTRODUCTION

Our colleagues, in the middle of next century, will know if we have taken the right decision at the appropriate time, or if we made a mistake, discontinuing long and regular series of observations without due care for a proper correlation with the modern techniques.

We must remember that if we go on with the classical observations longer than would be necessary, judging by the standards of the mid-twenty-first century, we do not lose anything of the valuable series of data accumulated till now. But if we discontinue our classical observations too soon, we shall introduce a big discontinuity in our records that cannot be recovered anymore, and our future colleagues will strongly critize us for taking this decision.

Our colleagues of about a hundred years ago were employing the first beginnings of what was later called statistics, and they had no idea about what are now described as correlation functions. Unfortunately, the word correlation is employed, even nowadays, in a loose sense and some scientists tend to forget that it has a precise statistical definition. Many observations are still presented in such a way that they do not conform with the theory of errors.

It is interesting to think that we are in the middle of a period of two hundred years at a time when new techniques are revolutionizing the acquisition of information about the Earth's rotation.

If the lessons of the past are applied to present circumstances, we should endeavour to set up central bureaus taking care of the observing programmes and adopting, as far as possible, the same type of instruments and the same observing techniques. The reduction of the observations should also be performed at the same computing centre in order to assure uniformity of the computations. The precisions obtained by these techniques are at least one order of magnitude greater than the classical techniques and, therefore, the problems of getting homogeneous series of observations are still more difficult than with the classical methods.

The debate, at the moment, is about which technique will be more useful or will give us more information at a lower cost. The adoption of one or more of the techniques mentioned will depend, to a great extent, on the amount of finance available for developing this type of research.

The possible experiments that might discriminate among the competing techniques have to be carefully planned, and we must not forget that the surface of our planet is subject to many forces, and that a more general model of the Earth should consider, for instance, its change of shape due to elasticity and the transference of heat. It is even pertinent to ask if there is such a thing as a fixed point on the Earth. The answer to this question depends, of course, on the definition of a fixed point and that, again, is related to the definitions of precision and accuracy of our measurements.

A good way to avoid all these difficulties is to set up, side by side, different techniques so we have no doubts about possible local displacements. The future will show us if we have adopted the right techniques and procedures.

The great advantages we have had with the functioning of the Central Bureau of the IPMS and the B.I.H. are widely recognised. In spite of the achievements obtained by these international organisations, we still suspect that some irregularities, in the pole path and in time determination, are due to some errors in the observing and computing programs.

We should forecast for the next decades the setting up of similar programmes of international cooperation dealing with observations made with Doppler, lasers to artificial satellites and the Moon, and VLBI. The results obtained by different techniques should be consistent, and that is another goal for the future.

Another colleague, speaking to a gathering dealing with Earth's rotation problems in a hundred years' time, will probably worry about new observational and computing techniques as we are doing today. Let us hope that our work will facilitate the job of our future colleagues of the next century.

PART I : TIME

BASIC PROBLEMS IN THE KINEMATICS OF THE ROTATION OF THE EARTH

Bernard Guinot
Bureau International de l'Heure
Paris, France

1. INTRODUCTION

With the advent of more precise methods for measuring Earth rotation, a number of corrections to the apparent directions in space, to the terrestrial references, and to the rotation axis motion have to be carefully applied. It is the duty of the international Astronomical Union to give recommended or conventional expressions of these corrections in order to avoid inextricable difficulties in discussing the evaluated results. However, this task is not sufficient. The concepts used in the description of the Earth's rotation are somewhat obscured by traditions. They should be purified by removing notions which are not directly relevant.

The following discussion will be restricted to the case where the observations are referred directly to a non-rotating frame given by the directions of stars (or of extragalactic sources; in the following, only the word "star" will be used, designating also these objects). The purpose of the discussion is to look for the minimum requirements for the kinetic study of the rotation of the Earth. It will be considered later how these requirements can be fulfilled without deviating too much from the usual practice in astrometry.

2. FUNDAMENTALS OF EARTH ROTATION MEASUREMENTS

To represent the Earth, a sphere with unit radius is used, where the points Z_i represent the directions of the plumb lines of classical instruments and of the base-lines of interferometers. We first assume that the Z_i have no relative motions, the sphere attached to them is called the <u>terrestrial sphere</u>.

Similarly, a sphere of unit radius can be attached to the star directions, represented by E_j, after correction for the proper motions and for the various effects due to the position of the observer, to his velocity, and to the atmosphere. This sphere is called the <u>non-rotating</u>

sphere.

Let us assume that that these two spheres are concentric. To describe the rotation of the Earth, it is sufficient to give their relative position as a function of time, t. This can be accomplished by giving the time-series of three parameters $\theta_i(t)$, i = 1, 2, 3, which can be chosen in an infinite number of ways.

This description immediately suggests a method of measurement of the Earth rotation: one has to observe, at successive instants t_1, t_2, ..., the positions of the Z_i among the E_j in order to get $\theta_i(t_1)$, $\theta_i(t_2)$, ... The only condition is that the instants be sufficiently close so that the $\theta_i(t)$ can be interpolated. Such observations are indeed routinely realized with classical instruments, and there is no need to refer to the rotation axis to reduce them, as soon as a definition of the three θ_i is chosen. The results would therefore be completely independent of the errors which may affect the precession and nutation.

However, for dynamical studies, it becomes necessary to refer to the position of the instantaneous axis of rotation, or to the angular momentum axis, or to the figure axis, or to any other axis with a dynamical definition, such as the one proposed by Atkinson (1973). The choice between these axes is only a matter of convenience. The word "pole" will designate the representative point on the spheres of the direction of one of these axes, towards the North.

The principles of the traditional method for measuring the Earth rotation are then as follows:

(a) DATA ASSUMED TO BE KNOWN

The motion of the pole on the non-rotating sphere (i.e. among the E_j) is described by the series A(t), B(t) of the luni-solar precession and nutation.

(b) OBSERVATIONS

The observations locate the terrestrial sphere with respect to the non-rotating sphere (i.e. the Z_i among the E_j) at an instant t_k.

(c) EVALUATION

The position of the pole on the terrestrial sphere (i.e. among the Z_j) is then derived in form of $u(t_k)$, $v(t_k)$, as well as the angular position of the terrestrial sphere, around the polar axis, $\theta(t_k)$.

It is now necessary to give 5 parameters, which is the minimum number of parameters to fix the relative position of the two spheres and one direction relative to them. In current practice, $\theta(t_k)$ is referred to the true equinox of date and the motion of this point appears as a sixth series. But the reference to the moving equinox is quite unnecessary, as we shall see later, and is an example of the complexities due to tradition. A more convenient origin can be chosen on the equator.

In this 5-parameter scheme, the series $A(t)$, $B(t)$ are affected by errors coming from two sources:

- the pole position on the non-rotating sphere at an initial date, t_o, is not perfectly known (for instance, using the present practice, the FK4 pole in 1950.0 does not coincide with the real position of the mean pole at this date):

- the conventional development of the luni-solar precession and nutation from t_o to t contains errors.

Therefore, even when neglecting all other sources of errors, $u(t)$, $v(t)$, $\theta(t)$ are also erroneous. However, the orientation of the Earth in space is restored correctly by the use of the values of the five parameters at a date, t, providing that the computations be made rigorously and that these parameters be given with a time resolution such that the shortest terms do not average out (time resolution of a few hours). Consequently, it could serve the interests of the users to tabulate in a single document, not only u, v and θ as functions of time, but also the precession-nutation series A and B employed to derive them.

This property is important in the cases where only a precise orientation of the Earth is needed. The definition of the pole and its location on the non-rotating sphere is immaterial, at least in theory. It is nevertheless useful, for practical reasons and for dynamical studies, that all the motions of the pole which can be modeled be taken into account (See Appendix I).

As already stated, all these considerations are only valid when the Z_i have no relative motions. If they have non-negligible relative motions they must be corrected for. The terrestrial sphere must then be attached to the Z_i in "a prescribed way" which has some arbitrary character. Similarly the relative motions of the apparent places of the E_j must be corrected. These corrections are not trivial when an accuracy of $0''.001$ is to be ensured. Appendix I lists some of them and points out some difficulties encountered in expressing their precise values.

3. THE NON-ROTATING REFERENCE SYSTEM

A non-rotating reference system is realized by a great circle and a reference point σ_o on it (Figure 1). This system is fixed among the directions of the stars and corrected for proper motions. The pole, S_o, of the circle defines the direction OZ_o; OX_o is along $O\sigma_o$; and OY_o completes the direction triad. The motion of the pole P will be described by the time series, $d = S_oOP$, and $E = \sigma_oS_oP$.

The instantaneous system OXYZ will be chosen so that OZ is along OP. The condition is imposed, that, when P moves on the non-rotating sphere,

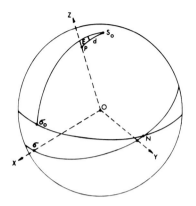

Figure 1. Non-rotating reference system.

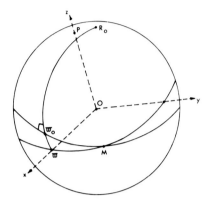

Figure 2. Terrestrial reference system.

the instantaneous system has no component of rotation around OZ. Let σ be the point where OX intersects the sphere. To realize the above condition it would be necessary to fix an origin on the equator at a date, t_o, then to derive σ at a date t, taking into account the whole history of the motion of P between t_o and t. Practically σ would be obtained by giving the quantities $\sigma_o N$ and $\sigma N - \sigma_o N$, N being the ascending node of the equator in the great circle of pole S_o. These quantities depend on E and d only. The nutation terms being small, the computations can be done for the precession only, and σN can be used on the true equator instead of the mean equator. σ will be called the <u>non-rotating origin</u>. One of the coordinates of a star is the usual declination, but the other differs from the conventional right ascension; it will be called the instantaneous ascension.

Atkinson has already proposed to use on the true equator a reference point freed from the nutation. In the above system, it is proposed to remove also the effects of luni-solar and planetary precessions.

4. THE TERRESTRIAL SYSTEM OF REFERENCE

The terrestrial system of reference is similarly defined by a great circle and a reference point $\overline{\omega}_o$ on it (Figure 2). The pole, R_o, of the circle gives the direction Oz_o; Ox_o is along $O\overline{\omega}_o$; and Oy_o completes the direction triad. This system is realized by the coordinates attributed to the Z_i and their variations. As the secular motion of P is slow, and the periodic components are small, it is possible to locate R_o so that P keeps close to it.

The instantaneous system has its Oz axis along the direction of the moving pole P. In order to cancel any component of rotation around Oz, the reference point $\overline{\omega}$ in the equator is obtained by $\overline{\omega}M = \overline{\omega}_0 M$, where M is the ascending node of the equator in the circle of the reference plane $x_0 O y_0$; Ox is along $O\overline{\omega}$. $\overline{\omega}$ is equivalently defined by the intersection of $R_0\,\overline{\omega}_0$ with the equator.

The present practice follows the above principle. For instance the initial latitudes and longitudes of the BIH, and their variable systematic corrections materialize a system of reference attached in a prescribed way to the Z_i. However, the relative positions of the Z_i are poorly known and the position of $\overline{\omega}_0$ is very uncertain. The consequence is that precise geodetic networks are obtained with undetermined rotations, especially around Oz_0.

5. THE ROTATION OF THE EARTH IN THE NON-ROTATING REFERENCE SYSTEM

We already defined the parameters

• E and d, which describe the motion of P on the non-rotating sphere,
• u and v, which describe the motion of P on the terrestrial sphere (the usual coordinates of the pole).

It remains to define the angular position around OP. This is obtained by the angle $\overline{\omega} O\,\sigma$, reckoned positive westward, which will be called the stellar angle and denoted by θ. Following its definition, θ expresses the sidereal rotation of the Earth, and its time derivative gives directly the angular velocity of the Earth in space.

As θ increases linearly if the rotation of the Earth is uniform, UT1 has the form

$$\text{UT1} = k(\theta - \theta_0),$$

the coefficient k being a constant chosen so that a day of UT1 is close to the duration of the mean solar day. Appendix II gives the expression of UT1 according to the above principles, with the present system of constants.

6. DISCUSSION

Before reaching the conclusions, additional comments will be made on the non-rotating reference system and the definition of UT1.

6.1. On the non-rotating reference system

A catalogue of stars defines its own pole and equinox, which do not coincide, in general, with the real pole and equinox. Although small, the departure of the catalog pole from the real pole is significant (it gives rise, in particular, to systematic errors, $\Delta \delta_\alpha$). The separation of the equinoxes may be large. In other terms, the accuracy defect of the practical definition of the equatorial system is larger than the internal defects of the catalog.

This unhappy situation cannot be expected to improve, particularly for the equinox. Already existing methods of classical astrometry, such as studies of the data of time and latitude services, locate the relative positions of the stars and the pole position fairly well, but they do not locate the equinox. The project of space astrometry (European Space Agency, 1976) does not locate the pole nor the equinox. Interferometry does not locate the equinox. It is therefore illogical to attach the whole coherent system to poorly measured directions and to rotate this system from time to time, to adjust it to the last evaluations of these directions.

The proposed system, definitively linked to the directions of stars, would be especially useful in the future, when the internal coherence of the catalogs will be improved. It also has the advantage of emphasizing the experimental character of the figures which express the position of the celestial pole and also of the pole of the ecliptic.

It can be considered that the pole and equinox of the future FK5 will be for some time the best estimates of the positions of these points among the fundamental stars. But it will not remain so, and it is suggested that corrections for these points be issued instead of rotating the whole system.

6.2. On the definition of UT1

UT1 has presently no clear definition. It is often presented as a form of the mean solar time, which has no theoretical nor operational advantages, but shows a discouraging complexity. In fact, only the sidereal rotation of the Earth has to be known, and UT1 could be merely cancelled, retaining only the stellar angle defined above.

As it would be unrealistic (on account of its broad use and relation to the UTC system) and somewhat unpractical (on account of the rapid variation of the stellar angle with respect to TAI) to abandon UT1, its definition as a function of the stellar angle is proposed in the conclusions.

7. CONCLUSIONS, RECOMMENDATIONS

From the considerations listed in this paper result the following proposals.

7.1. Symposium No. 82 should indicate what is the <u>definition of the pole</u> which seems suitable (pole of rotation, pole of angular momentum, pole proposed by Atkinson, ...).

7.2. The Symposium should adopt the <u>expressions of a few corrections</u> for effects entering in the study of the rotation of the Earth, which are fairly well known and recommend their general use (Earth tides, diurnal nutation, relativistic deflection of light, variation of UT1 due to zonal tides). Furthermore, accurate methods of reduction should be recommended, the aim being uncertainties smaller than 0".001.

7.3. Concerning the <u>definition of the space reference system</u>, it is suggested

(a) that this system be attached to the directions of stars or extragalactic sources without reference to the position of the pole and of the equinox of some date,

(b) that the coordinates of the pole of rotation be given in this system by two time series,

(c) that, similarly, the coordinates of the pole of ecliptic be given in this system.

The four parameters stated in (b) and (c) contain all the basic information for fundamental astrometry.

7.4. Concerning the <u>realization of the space reference system</u>, it is suggested

(a) that, for the near future, the FK5 be considered as the official realization of the system mentioned in 7.3(a), and that, at its standard date, its pole and its equinox (which, together with the conventional value of the inclination of the ecliptic, give the pole of ecliptic) be the basis of the coordinates stated in 7.3(b) and 7.3(c);

(b) that, later, no attempt be made to adjust the pole and equinox of the FK5 to the observed positions of these directions; the improvement of the FK5 should consist of

- the reduction of internal inconsistencies in positions and proper motions at the standard epoch,
- a rotation rate affecting the proper motions, if required, in order to link it to a better observed non-rotating reference system (for instance linked to extragalactic sources);

(c) that, if a significant departure of the observed pole and equinox from the pole and equinox of the FK5 is found, the position and motion of these directions in the FK5 system should be given, without rotating the FK5 system.

7.5. Concerning the <u>instantaneous equatorial reference system</u>, it is recommended that a non-rotating origin be adopted on the instantaneous equator, instead of the conventional equinox. This point should coincide with the mean equinox of the FK5 at its standard epoch.

7.6. Concerning the definitions of Universal Time:

(a) The hour angle of the non-rotating origin from the prime meridian, here denoted by stellar angle, should be the basis of the definition of UT1. A self contained definition of UT1 could be
"UT1 is an angle which is proportional to the sidereal rotation of the Earth, the coefficient of proportionality being chosen so that UT1, in the long term, remains in phase with the alternation of day and night. In some applications, UT1 can be considered as a non-uniform time scale."

(b) The definition of UT2 should be abolished.

(c) It could be useful to define a new form of UT, in which the variations due to zonal tides are removed.

7.7. Concerning the orientation of the Earth in space, in order to provide users with all the data they need in a single document, it is recommended that the services dealing with the rotation of the Earth tabulate not only UT1-TAI and the coordinates of the pole in the Earth linked system, but also the coordinates of the pole in the space reference system, which are used in the reduction of the observations.

APPENDIX I

CORRECTIONS TO THE TERRESTRIAL AND APPARENT SPACE REFERENCES AND TO THE ROTATION OF THE EARTH

We will deal in this Appendix with the various corrections which are needed to remove the relative motions of the terrestrial references and of the space references as seen from the Earth. We will also consider the terms of the rotation of the Earth which can be modeled with an accuracy matching the precision of the measurements.

In classical astrometry the final precision of the measurements of Earth rotation is of the order of $0\overset{''}{.}01$. It is thus satisfactory that the corrections be known with uncertainties of a few $0\overset{''}{.}001$. The recent decisions taken by the International Astronomical Union meet this requirement (but the relative positions of the stars and of the zeniths are not known to this level of precision except for Earth tides and for some relativistic effects).

With the advent of methods which could lead to an observational precision of $0\overset{''}{.}001$, or better, what appeared previously as known corrections becomes a subject of study, and there is little that can be done presently. There are, however, a few terms for which the current practice can be improved, and only these terms will be considered here.

1. Corrections for relative motions of the terrestrial references

In order to derive the <u>corrections due to Earth tides</u>, it was recommended that clinometers, gravimeters, strain-meters..., be placed in the vicinity of astronomical instruments. But this recommendation had little effect. This failure can probably be explained by the difficulty of interpretation of the data. On the other hand, it is not always felt necessary to have local measurements. A possible issue could be to recommend that, in the absence of reliable measured effects of the Earth tides, conventional corrections be applied with adopted values of h, k, ℓ, without phase lag, or local corrections. It is too early to adopt a <u>model of plate motions</u>.

2. Corrections for the apparent relative motions of celestial sources

The complete <u>correction for proper motions</u> of stars would require the knowledge of the ratio of radial velocity/distance, in order to take care of the foreshortening effect. In some cases, this effect is not negligible.

The <u>aberration</u> should be computed so that the geometric direction is obtained by adding to a unit-vector along the apparent direction, the vector, $-\vec{V}/c$ ($\vec{V}$ is the velocity of the observer, c the velocity of light). The reverse method (unit-vector along the geometric direction) is often used. The difference between the two methods amounts to $0\rlap{.}''001$ for the annual aberration. It should also be remembered that the component of the velocity of the Earth perpendicular to the ecliptic and the perturbations in longitude give rise to aberrations of the order of $0\rlap{.}''001$.

In classical astrometry, the <u>relativistic deflection of light</u> in the gravitational field of the Sun is usually omitted. However, even for night observations, this deflection is of the order of a few $0\rlap{.}''001$. This effect has been studied by Brandt (1974). Kimura (1935) already attempted to apply this relativistic correction to the data of the International Latitude Service. Interferometric measurements should be also corrected for the effects of the gravitational field of the Sun, in addition to the relativistic effects which appear in the computation of the time delay and fringe rate (Thomas, 1974).

3. Corrections for the rotation of the Earth

3.1. Definition of the pole and motion of the pole

Following the proposal by Atkinson (1973) to replace the instantaneous pole of rotation, which is the present axis of reference, by a new axis freed from the forced diurnal nutation with respect to the Earth, many discussions within the IAU led to contradictory recommendations. We will not comment on the possible choices in this paper, but only recall that, according to the definition of the pole, some terms appear either in the motion of the pole on the terrestrial sphere or on the non-rotating sphere. Anyway, the forced diurnal nutation has to be taken into account. The elasticity of the Earth changes only slightly its amplitude, as shown theoretically by McClure (1973). This nutation has been found in the observations (McCarthy, 1976).

Besides the precession-nutation, for which the IAU has prepared improved numerical coefficients, and the "sway" which can reach 0!'001, no other motions of the pole can be modeled sufficiently well for the computations.

3.2. Variations of UT1 due to zonal tides

These variations (Woolard, 1959; Pil'nik, 1970) are real irregularities of the rotation of the Earth; in principle, they should require no correction. However, the terms with the shortest periods may cause some difficulties in the interpretation of measured values of UT1, if the time resolution is not sufficient. For instance, the terms with periods 13.7 d and 27 d have amplitudes of the order of 0.8 ms; they produce problems in handling the 5-day averages of UT1 determined by the BIH with random uncertainties (1σ) of 1 ms. In addition, terms with long periods have large amplitudes (0.15 s for the term of 18.6 years) and they must be removed before investigating the unmodeled variations of UT1.

Several of the short period terms have been experimentally found by many authors, in particular by Pil'nik (1970). It was observed that the 13.7 d term had an amplitude which did not correspond to the theory (Guinot, 1974); but subsequent studies (Guinot, not published) indicate that the discrepancy might be due to an additional term with a period of 13.70 d of unknown origin.

It might be advisable to define a form of UT (UT3 ?) which would be UT1 corrected for the zonal tide effects. If this suggestion is accepted, the conventional expression of the correction should be given.

3.3. Suppression of UT2

The seasonal variation of UT1 is variable from year to year (see, for instance: Okazaki, 1975, 1977; Lambeck and Cazenave, 1973). The conventional value of UT2, obtained from UT1 by addition of periodic annual and semi-annual components is therefore of little use. It is proposed to cancel the definition of UT2.

APPENDIX II

DEFINITION OF UT1

The present definition of UT1 is contained in the well known expression giving the mean sidereal time at 0h UT:

$$T = 6^h\ 38^m\ 45\overset{s}{.}836 + 8640\ 184\overset{s}{.}542\ t + 0\overset{s}{.}0929\ t^2,$$

t being measured in Julian centuries of 36525 days of UT1 from 1900 January 0, 12h UT1.

We assume that the non-rotating reference system is the mean equatorial system on 1900 January 0, 12h UT1. On the mean equator of date, the non-rotating origin is then derived from the mean equinox of date by a rotation $\zeta_A + z_A - s$ eastward, where s is a small quantity integrated along the path of the mean pole. Therefore the stellar angle is

$$\theta = T - \zeta_A - z_A + s.$$

With the present system of constants

$$\zeta_A + z_A = 46\ 085\overset{''}{.}06\ \tau + 139\overset{''}{.}73\ \tau^2 + 36\overset{''}{.}32\ \tau^3$$

$$s = 36\overset{''}{.}28\ \tau^3 - 0\overset{''}{.}04\ \tau^4$$

τ being reckoned from 1900.0, in units of 1000 tropical years. Expressing $\zeta_A + z_A$, with t as unit, disregarding the negligible effects of the non identical origins for t and τ and for the irregularity of the Earth rotation,

$$\zeta_A + z_A - s = 307\overset{s}{.}2403\ t + 0\overset{s}{.}0932\ t^2.$$

Thus, the expression of θ at 0h UT is

$$\theta = 6^h\ 38^m\ 45\overset{s}{.}836 + 8639\ 877\overset{s}{.}302\ t - 0\overset{s}{.}0003\ t^2.$$

The small term in t^2 results from small inconsistencies in the system of constants. Omitting this term, the general expression of UT1 has the form

$$UT1 = k(\theta - \theta_0)$$

given section 5.

This definition of UT1 leads to the same value of UT1 as the usual expression, if the instantaneous ascension is employed instead of the usual right ascension.

REFERENCES

Atkinson, R. d'E.: 1973, Astron. J., 78, p. 147.
Brandt, V. E.: 1974, Astron. Zh., 51, p. 1100.
European Space Agency, 1976, Space Astrometry, Report on the Mission Definition Study, DP/PS (76) 11.
Guinot, B.: 1974, Astron. and Astrophys., 36, p. 1.
Kimura, H.: 1935, Results of the Int. Latitude Service 7, p. 39.

Lambeck, K., Cazenave, A.: 1973, Geophys. J. R. Astron. Soc. 32, p. 79.
Lieske, J. H., Lederle, T., Fricke, W., Morando, B.: 1977, Astron. Astrophys. 58, p. 1.
McCarthy, D. D.: 1976, Astron. J. 81, p. 482.
McClure, P.: 1973, "Diurnal Polar Motion", Goddard Space Flight Center Report X-592-73-259.
Okazaki, S.: 1975, Publ. Astron. Soc. Japan 27, p. 367.
Okazaki, S.: 1977, Publ. Astron. Soc. Japan 29, p. 619.
Pil'nik, G. P.: 1970, Astron. Zh. 47, p. 1308.
Thomas, J. B.: 1974, Proc. 6th Annual PTTI, p. 425.
Woolard, E. W.: 1959, Astron. J. 64, p. 140.

DISCUSSION

F.P. Fedorov: Do you propose that the definition of the pole that was adopted in Kiev a year ago, specified by the instantaneous axis, be replaced by Atkinson's definition?

B. Guinot: The instantaneous rotation pole is not especially convenient for the reduction of astronomical observations, or for theoretical work. Although I did not specify in my paper which point should be adopted for the terrestrial and celestial pole, I believe now that Atkinson's pole is the best.

A NEW STATE TIME AND FREQUENCY STANDARD OF THE USSR

S. Pushkin
USSR Gosstandart
Moscow, U.S.S.R.

ABSTRACT

In 1976 a new state primary time and frequency standard of the USSR was certified and confirmed with an error of reproducibility less than 1×10^{-13} and with an unavoidable systematic error less than $3-4 \times 10^{-13}$. This standard includes a laboratory primary cesium beam frequency standard providing an independent definition of units in the SI system, primary hydrogen frequency standards preserving the units of frequency and time intervals, and hydrogen and cesium clocks providing the TA(SU) and UTC(SU) time scales.

Measurements within the standard are made automatically by computer. Comparisons within the Soviet Union and with standards of other countries are made by television, portable atomic clocks, and meteor trails. Errors in comparisons are less than $0.1 - 0.5$ μs. Loran-C is used as a reserve means of comparison.

1. INTRODUCTION

The USSR state primary time and frequency standard is the basic means for the measurement of time and frequency in the Soviet Union. This standard includes systems which produce the units of frequency and time interval, provide continuous USSR astronomical and coordinated time scales, and provide a means of comparison with external and internal standards. Secondary systems guarantee an uninterrupted power supply and provide for measurements of environmental and system parameters. The basic elements of the state standard are the systems for the reproducibility of frequency and time interval as well as the TA(SU) and UTC(SU) time scales.

2. CESIUM BEAM FREQUENCY STANDARD

The cesium beam frequency standard developed in 1975 provides frequency reproducibility of the unperturbed transition of the cesium 133 atom to the order of $3-4 \times 10^{-13}$. Comparisons of this standard are made with the hydrogen frequency standards included in the state standard.

3. HYDROGEN FREQUENCY STANDARDS

The high stability and reproducibility of the hydrogen frequency standards permit almost a half order-of-magnitude decrease in the uncertainty in the length of the second in the state standard. The hydrogen standards were first incorporated in the state standard in 1967. In 1976 hydrogen standards of a new design, characterized by an error in reproducibility of the order of $0.6 - 1.2 \times 10^{-13}$, were introduced in the state standard. Three sets of hydrogen standards, whose frequencies are periodically defined by comparison to the cesium frequency standard, are incorporated in the state standard.

4. TIME SCALE

The clock time scale is produced by three hydrogen and six cesium clocks. The gradual transition from cesium to hydrogen clocks is expected because of the smaller variation in the frequency of the hydrogen clocks. This will occur as the operational reliability of the hydrogen clocks improves. The TA(SU) and UTC(SU) time scales are calculated analytically through the intercomparison of standards by an electronic computer. A working time scale $UTC(SU)_w$ is maintained to within ± 0.2 μs from UTC(SU).

5. INTERNAL COMPARISONS

All of the time and frequency measurements are made automatically and recorded on magnetic tape, which is then processed by electronic computer. The error of a time comparison is ± 1 ns while the error of a frequency comparison is of the order of $\pm 1 \times 10^{-14}/1000$ s.

6. EXTERNAL COMPARISONS

Comparison with external standards is accomplished by various techniques. Time and frequency comparisons are made using VLF and LF radio signals. The uncertainty is less than $0.5 - 3.0$ μs. Television signals are also used for comparison. For a distance of 500 km the uncertainty of a measurement is less than 0.5 μs. Meteor trail reflections may also be used with an uncertainty of $0.2 - 0.3$ μs. The most accurate method is the use of portable atomic clocks with an uncertainty of $0.5 - 0.10$ μs. Comparisons of UTC(SU) with UTC(BIH) have been carried out using portable clocks and Loran-C.

7. CONCLUSION

The state primary standard has the following metrological characteristics: a systematic error less than $\pm 4 \times 10^{-13}$, an error of reproducibility of frequency and time intervals less than 1×10^{-13}. Work is currently in progress to decrease the error of frequency reproducibility of the cesium beam frequency standard to 1×10^{-13} and that of the hydrogen standards to $2 - 3 \times 10^{-13}$. Efforts are also being made to improve the automization of comparisons and to develop new types of clocks. Work on the use of lasers is now in progress. The He - Ne laser (λ = 3.39 µm) is characterized by an error in frequency reproducibility of the order of 6×10^{-13}. Direct comparisons of the laser frequency to the primary state standard frequency and to a D_2O laser (λ = 84 µm) frequency are made with a comparison uncertainty of 1×10^{-13}. Investigation of ways of improving the accuracy of lasers is continuing.

A NEW METHOD OF UNIVERSAL TIME COMPUTATION USED IN THE STATE TIME AND FREQUENCY SERVICE OF THE USSR

D. Yu. Belotserkovskij and M. B. Kaufman
Inst. of Phis.-Tecn. Measurements
Moscow, USSR

Processing of astronomical observational data of several time services and the publication of Universal Time of emission of time signals has been carried out in the USSR continuously for fifty years. Currently the Universal Time computations are made in the USSR State Time and Frequency Service using the observational data of eleven Soviet time services and ten time services of socialist countries that have volunteered to participate. The functioning of this system simultaneously with the BIH system that unites time services from all over the world has certain practical advantages. The Soviet Universal Time scale UT1(SU) may be influenced by some geophysical effects which are characteristic of the Eurasian continent, and consequently, it may be used advantageously for scientific and practical purposes in this region. The agreement between different time scales, produced by various methods and observational data, also makes it possible to evaluate the reliability of these scales.

Beginning in 1928, Universal Time computations have been made in the USSR by the "differential" method proposed by Prejpich (1933). This method is characterized by preliminary smoothing of the observational results of each contributing service. The BIH has used a similar method since 1931 when observational data from different time services were organized (Stoyko, 1946). The improvement of timekeeping methods and radio reception facilities made it possible to systematize astronomical observational data processing, to compare the data with one master clock, and to smoothe the combined data obtained by all time services. This method conceived in the USSR by N. N. Pavlov was put into practice by D. Yu. Belotserkovskij in 1951 (Belotserkovskij, 1967).

With the appearance of frequency standards and uniform atomic time scales, Universal Time remained in use only as a measure of the Earth's rotation (rotational time). Now there is no need for complete smoothing of astronomical observational data, but it is desirable that the information on the non-uniformity of the Earth's rotation be incorporated into the Universal Time scale. A new method of processing observational data has been utilized by the USSR State Time and Frequency Service since 1975 (Kaufman, 1976)

A probability-statistical approach is the main feature of this method which is analogous to the well-known method of smoothing proposed by Whittaker (Whittaker et al., 1960) and modified by Vondrak (1969). However, due to the necessity of processing a combination of statistically non-uniform data series obtained by different time services, different observers and instruments, the main condition for smoothing is enlarged by terms that take into account systematic observational errors, δ, and their variations with time. Thus we have the equation of condition,

$$\mu^2 \sum_j (\Delta''' u_j)^2 + \sum_i \lambda_i^2 (\delta_i - \delta_i')^2 + \sum_{i,j} P_{ij} (u_{ij}^* - \delta_i - u_j)^2 = \min., \quad (1)$$

where i refers to the "observer-instrument" combination; j the date of observation; $u_j = (UT1 - UTC)_j$ the value of the difference between the rotational time and the coordinated time at 12^h UT for the date, j; $\Delta''' u_j$ the third difference of successive u_j values; $u_{ij}^* = UT1_{ij}^* - UTC_j$ the observational results referred to the CIO; P_{ij} the weight of the observations characterizing their accuracy and depending on random errors; λ_i^2 coefficients characterizing the degree of systematic error (stability); and μ^2 the parameter characterizing the degree of smoothing for the series of u_j.

Universal Time computations are made weekly using a thirteen-day data interval from the preceeding Friday through the following Wednesday. This presumes that the systematic errors do not change during this thirteen-day interval and that the combination of the variations $\delta_i - \delta_i'$ between two neighboring intervals for all observers and instruments are random.

Thus the conditional equation (1) is a combination of three inter-related conditions for minimum sums of squares of three groups of random quantities: third differences $\Delta''' u_j$; differences of systematic errors, $\delta_i - \delta_i'$; and random errors, $u_{ij}^* - \delta_i - u_j$. The first of these conditions ensures a desirable degree of smoothness of the series of u_j; the second ensures the closeness of the systematic errors δ_i to their values δ_i' during the preceding interval (preservation of the system of the UT1 time scale); the third ensures the best agreement of the unknown quantities u_j and δ_i with the observations (suppression of random errors). The main feature of this method is the fact that no supposition is made beforehand as to the functional form of successive values for either the Universal Time or the systematic errors to be found. Only statistical limits are imposed on their limits.

The computation of the unknowns, u_j and δ_i, and their mean errors, m_{u_j} and m_{δ_i} is usually made by means of the method of least squares in which the system of linear equations resulting from the conditional equation (1) are solved. The number of such equations is equal to the number of "observer-instrument" combinations plus thirteen (the number of days to be processed). Calculated values of δ_i are used to replace the δ_i' in the processing of the next interval.

The weights, p_{ij} and the coefficients λ_i^2 are calculated by

$$p_{ij} = \frac{1}{m_i^2} (n_{ij})^{1/2}, \quad \text{and} \quad \lambda_i^2 = \frac{1}{m_{\delta i}^2 + \sigma_i^2},$$

where n_{ij} is the number of individual time observations (groups of stars) for which a value of u_{ij}^* is determined during one evening, m_i the mean error for one observation of time, σ_i the mean square value of the variations $\delta_i - \delta_i'$ between neighboring intervals, and $m_{\delta i}$ the mean error of δ_i'. Values of m_i and σ_i are determined by analysis of the results for each "observer-instrument" combination of the preceding calendar year. The parameter μ^2 is calculated by

$$\mu^2 = \frac{1}{26} \sum_{i,j} p_{ij},$$

which ensures the stability of the frequency response characteristics of the observational smoothing for different intervals. This method is characterized by a frequency response which can distinguish periodic variations having periods greater than seven days (Figure 1). For comparison the frequency response characteristics of the Pavlov method and that of the BIH method (Annual Report, 1970) are also displayed in Figure 1.

However, the effective resolution of this method is obtained at the expense of an increase in the mean error for daily values of Universal Time which is estimated to be 2 - 5 ms. Of course, if these values were averaged over five-day intervals it would be possible to diminish these estimates to 1 - 2 ms. In that case one would smoothe real, short-period fluctuations (in particular, fortnightly variations). This is confimed by spectral analysis of the data

In a plot of spectral density (Figure 2) for daily values of UT1 - UTC calculated by our method some peaks are distinguished. These peaks represent short-period, non-uniformities in the Earth's rotation and fluctuations of verticals. The heights of the tidal peaks are in good agreement with theoretical values (shown by circles). Also shown is a plot of spectral density of raw five-day BIH values of UT1 - UTC (Feissel, 1976) which shows that the M_f wave is noticeably decreased in comparison to the M_m wave.

The Universal Time data computed by the new method have been published since 1975 in weekly bulletins (Series A) and every three months in Series E bulletins. Series A bulletins present preliminary data calculated from current observations using preliminary polar coordinates from the BIH. Once a month Series A bulletins list the final UT1(SU) - UTC that have been computed using the polar coordinates given in BIH Circular D. Series E bulletins contain the Universal Time data for a

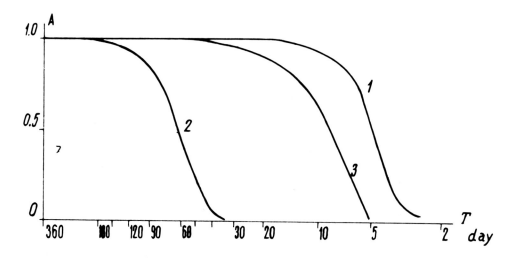

Figure 1. Frequency response of 1) this paper, 2) Pavlov's method, and 3) BIH method.

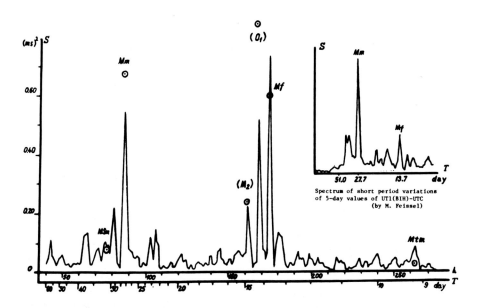

Figure 2. Spectrum of short period variations of daily values of UT1(SU) - UTC.

three-month period and tables of observational astronomical data. We are convinced that the publication of these data should be the duty of any bureau engaged in the work of centralized processing of astronomical observations. Only in this way will it be possible for skilled persons to obtain additional information from the observational data.

REFERENCES

Annual Report of the BIH: 1970, Paris.
Belotserkovskij, D. Yu.: 1962, Trudy Institutov Komiteta (Standartov) 58, p. 118.
Feissel, M.: 1976, in "Proceedings of the Eighth Annual PTTI Applications and Planning Meeting", Goddard Space Flight Center Publ. X-814-77-149, Greenbelt, Maryland, pp. 47-54.
Kaufman, M. B.: 1976, Trudy VNIIFTRI 29, p. 59.
Prejpich, N. Kh.: 1933, Trudy VIMS 3, p. 19.
Stokyo, N.: 1946, Bulletin Horaire Serie D, Janvier-Juin.
Vondrak J.: 1969, Bull. Astron. Inst. Czech. 20, p. 349.
Whittaker, E., and Robinson, G.: 1960, "The Calculus of Observations", London.

DISCUSSION

B. Guinot: The amplitude of the M_f wave in UT1 is reduced by a factor of between 0.6 and 0.7 in the BIH 5-day means, not by 0.1. The discrepancy between theoretical and observed M_f response seems to be due to an additional term, close in frequency but of unknown origin.
Ya. S. Yatskiv: I agree.

ON THE DETERMINATION OF UT1 BY THE BIH AND THE U.S.S.R. TIME SERVICE

Ya. S. Yatskiv, A. A. Korsun', N. T. Mironov
Main Astronomical Observatory
Ukrainian Academy of Sciences
Kiev, U.S.S.R.

ABSTRACT

The reference systems for the determination of UT1 and polar coordinates are discussed. Some arguments are given in support of the adoption of the mean pole origin (MPO) defined by the mean latitudes of the observatories. Taking into account the frequency response of the filters used by the BIH and the U.S.S.R. Time Service, the effects of the errors of the adopted nutational coefficients and tidal variations of the vertical are estimated.

1. INTRODUCTION

The following astronomical systems of terrestrial coordinates are used in the study of the Earth's rotation:
MPO - The mean pole origin defined by the mean latitudes of observatories at any moment. For determining the mean latitudes different methods may be used, such as the well-known Orlov's method.
CIO - The Conventional International Origin defined by the fixed latitudes of the ILS stations.
BIH 1968 - The reference system adopted by the BIH and referred to the epoch of 1968.
UT(SU) 1975 - The reference system adopted by the U.S.S.R. Time Service and referred to the epoch of 1975.
Some information concerning the observational data and the methods used for the determination of the above reference systems are given in Table 1. The merit of any reference system for studying the rotation of the Earth depends on:
(1) the contribution of the secular polar motion (if it exists) to the secular variations of the mean latitudes and longitudes of the stations,
(2) the stability of the periodic, nonpolar variations of the latitudes and longitudes of the observing stations,
(3) the statistical properties of the errors of astronomical observations,
(4) the methods used for preserving the reference system.
Let us consider some of these problems.

Table 1. Astronomical reference systems of terrestrial coordinates.

Name of System	Number of Observatories	Number of Instruments		Interval of Observations	Primary Reference System
		Time	Latitude		
MPO	all	-	-	more than 1.6 yr.	instrumental
CIO	5	-	5	1900.0 - 1906.0	instrumental
BIH 1968	51	48	39	1964.0 - 1967.0	CIO
UT(SU) 1975	5	8	-	1957.0 - 1971.0 1974.0 - 1975.0	BIH 1968

2. SECULAR AND LONG-PERIOD POLAR MOTION

The existence of the secular and long-period variations of the mean longitudes and latitudes of stations has been proved by many authors. These variations may be both real and fictitious. The former are due to the motion of the Earth's rotational axis, crustal movements, and variations in the direction of gravity, etc. The latter may result from errors of astronomical observations and their reduction. Using the data of the five ILS stations, many determinations of the secular polar motion have been carried out (Yumi and Wako, 1960; Mikhailov, 1971). It has been shown that the parameters of the secular polar motion vary considerably in time and depend on the combination of the ILS stations. On the average, for the time interval from 1900 to 1968 the mean pole has been moving with respect to the CIO with the velocity $0\overset{''}{.}0030$ - $0\overset{''}{.}0040$ per year in the direction of the meridian $73°$ W. The observed variation, B_o, and the theoretical variation, B_t, of the mean latitudes of several stations are given in Table 2.

Table 2. Linear trends of mean latitudes.

Station	Instrument	Interval of Observations	B_o / B_t (in $0\overset{''}{.}001$ / year)
Mizusawa	VZT	1900 - 1972	-3.0 / -2.1
Mizusawa	FZT	1940 - 1972	+4.0 / -2.1
Mizusawa	PZT	1957 - 1972	-3.0 / -2.1
Tashkent	VZT, T1	1895 - 1896 1969 - 1970	-4.2 / -3.4
Washington	PZT	1915 - 1972	+3.0 / +3.0
Richmond	PZT	1949 - 1972	+0.1 / +3.0
Ukiah	VZT	1900 - 1967	+2.7 / +2.4

These trends were calculated assuming that the secular polar motion based on the data of the five ILS stations was real. This table does not explicitly confirm the reality of the secular polar motion. Moreover, Fedorov (1975) showed that the values of the linear trends of the mean latitudes of the ILS stations does not contradict the hypothesis of the random character of these trends.

At present there is no way to resolve the problem of how much is secular polar motion and how much is due to other secular effects providing that the number of stations is limited, for example, to five. However, it is possible to solve the problem qualitatively. We have calculated the index, k, the sign of which might be used as a criterion for a decision on the reality of the secular polar motion derived from the ILS data. We let

$$k = (S_2^2/S_1^2) - 1,$$

where

$$S_1^2 = \frac{1}{n} \sum_{j=1}^{n} (\psi_j - \psi_{jo})^2, \quad S_2^2 = \frac{1}{n} \sum_{j=1}^{n} (\psi_j - \psi_{jo}) - (\Delta\psi_j - \Delta\psi_{jo})^2,$$

ψ_j is the yearly mean value of the latitude or longitude of the station, j, reckoned from an arbitrary initial value, ψ_{jo}, $\Delta\psi_j$ is the correction for secular polar motion derived from the ILS data and reckoned from an initial vaule, $\Delta\psi_{jo}$, and n is the number of stations. If the sign of k is negative it would support the reality of the secular polar motion. The results are given in Table 3. They do not support the reality of secular polar motion. We can conclude that the secular trend of the CIO relative to the MPO is, in the most part, due to the local nonpolar effects of the ILS stations.

Table 3. Values of the index, k.

	Latitude		Time	
Year	n	k	n	k
1955	10	+1.18	24	+0.20
1956	10	+0.93	25	+0.03
1957	17	+0.72	28	+0.08
1958	18	+0.79	28	+0.05
1959	21	+0.57	28	+0.01
1960	21	+0.20	32	0
1961	21	−0.14	32	0
1962	28	0	32	0
1963	28	0	32	+0.08
1964	27	0	32	+0.16
1965	26	+0.12	32	+0.01
1966	26	+0.19	−	−
1967	26	+0.26	−	−
1968	26	+0.19	−	−

If the secular polar motion does not exist, we should clear up whether there is some long-period motion with a period comparable to the interval of observation. Markowitz (1970) claimed to reveal the 24-year libration of the mean pole. Recently this problem was carefully studied by Vicente and Currie (1976). They have found a 30-year period in the polar coordinates. We have tried to find new support for this discovery, but failed. The polar coordinates have been calculated for the period 1905 - 1940 from observations of two groups of stations:
   Group 1 (x_1, y_1) - Mizusawa, Carloforte, Ukiah;

Group 2 (x_2, y_2) - Pulkovo, Greenwich, Washington.
The value of (x_1, y_1) and (x_2, y_2) have been filtered and analyzed by the same method as given in the paper by Vicente and Currie (Figure 1).

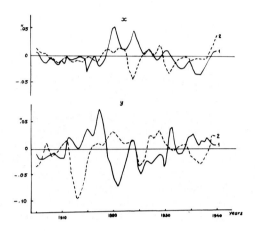

Figure 1. Long-period variations of polar coordinates derived from two groups of stations.

The significant periods found by the maximum entropy power spectrum analysis are 23.5 years and 22.0 years in x_1 and y_1, respectively, and 20.0 years and 29.5 years in x_2 and y_2, respectively. There are no correlations between the values of x_1 and x_2, y_1 and y_2. The long-period components in polar motion appear to result from local, nonpolar variations of the latitudes of some stations. We have concluded that the main contribution to these components is from the Ukiah station. Figure 2 shows the variation of the angle (S_{uk}) between the verticals of Ukiah and Kitab (Tschardjui) stations as well as the mean latitude of Ukiah (ϕ_u). There is a significant correlation between these two curves, the amplitudes of which are about 0″.20. The variations of S_{uk} do not depend on polar motion (Fedorov et al., 1972). Therefore the 20 to 30-year variation of mean latitude of Ukiah is nonpolar and may contribute to variations of x_1 and y_1 with an amplitude of about 0″.05. Figure 3 also gives strong support to this statement.

3. ON THE PRESERVATION OF THE REFERENCE SYSTEM BY THE BIH AND THE U.S. S.R. TIME SERVICE

To preserve the reference system the BIH uses fixed corrections to the data of the individual instruments to account for the local variations of latitude and longitude in the form of a constant, an annual and a semi-annual term:

$$R_j = a_j + b_j \sin 2\pi\theta + c_j \cos 2\pi\theta + d_j \sin 4\pi\theta + e_j \cos 4\pi\theta, \quad (1)$$

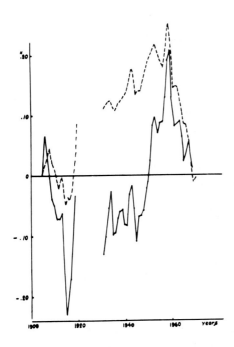

Figure 2. Variations of angle between the verticals of Ukiah and the Tschardjui-Kitab stations (solid line) and the mean latitude of Ukiah (dashed line).

where a_j, b_j, c_j, d_j, and e_j are the coefficients to be determined, and θ is the frequency of one cycle per year. It is assumed that R_j represents well enough the nonpolar variations of latitude and longitude of the stations. Clear tests are possible to check this assumption. We have calculated the angles between the verticals of several stations, S_O and S_C, based on the observational data as well as the coefficients given by the BIH, respectively.

Let D_O and D_C be the dispersions of the observed values S_O, and calculated values, S_C, respectively. In all cases we found $D_O \gg D_C$. Power spectrum analysis of S_O showed the existence of many periodic components besides the annual and semi-annual components. For example, we show in Figure 4 the power spectrum of variations of the angle between the verticals of Mizusawa and Paris from 1968 to 1976. The significant periods are 6.0, 1.0, 0.6, 0.5, and 0.4 years. The dispersion, $D_O = 2.095 \; (0\rlap{.}{''}1)^2$, is twice as large as D_C.

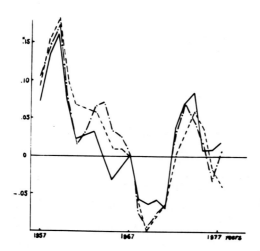

Figure 3. Variations of the angle between the verticals of stations and their mean latitudes. Solid line (S_{uk}) is the angle between the verticals of Ukiah and Kitab VZTs. Dashed line (S_{um}) is the angle between the verticals of Ukiah and mean observatory (Pulkovo, Kazan, Poltava, Kitab zenith telescopes). Dashed line with solid circles (ϕ_u) is the mean latitude of Ukiah with the opposite sign.

Therefore in our opinion the method for preserving the reference system by the BIH could be improved by using the statistical prediction of residuals, R_j, in the form,

$$R_{jt} = \sum_k P_{jk} R_{jt-k}, \qquad (2)$$

instead of the representation (1). The terms, P_{jk}, in (2) are the coefficients of the prediction error filter. In the computation process of the U.S.S.R. Time Service for preserving the reference system the following condition is used:

$$\sum_j \lambda_j^2 (\delta_j - \delta_j')^2 = \text{minimum}. \qquad (3)$$

The terms, δ_j and δ_j', are the systematic errors of time observations by the instrument, j, for two successive intervals of time. This condition corresponds to the assumption that the values of δ_j can be described by the Markov random process. This is not the case, and the use of (2) would be preferable.

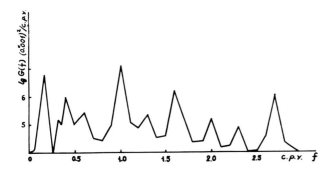

Figure 4. Power spectrum of S_o for Mizusawa and Paris stations.

4. SYSTEMATIC ERRORS OF TIME OBSERVATIONS

4.1. Correction for the errors of nutation coefficients and the forced diurnal motion of the pole

Currently nutation is computed from a theory of the rotation of the Earth as a rigid body and only the coefficient of the principal term is determined from observations. If the values of the coefficients in obliquity, a_i, and in longitude, b_i, are in error, the values of the clock correction will contain errors depending on both the right ascension, α, and declination, δ, of the star as well as the coordinates of the perturbing bodies such as the longitude of the lunar node, Ω. These errors can be represented by

$$\Delta u_i = \tfrac{1}{2}(\Delta a_i + \Delta b_i)\cos(\theta_i t + \beta_i - \alpha)\tan\delta$$
$$+ \tfrac{1}{2}(\Delta a_i - \Delta b_i)\cos(\theta_i t + \beta_i + \alpha)\tan\delta \quad (4)$$

where θ_i and β_i are the frequency and the phase of the corresponding nutation terms, and Δa_i and Δb_i are the corrections to the adopted nutation coefficients. Variations of longitude due to the effect of the forced diurnal motion of the pole are of the same form as (4). For this reason there is no possibility of deriving these effects separately from the observational data.

The BIH makes the corrections for the forced diurnal motion of the pole predicted by the theory of the rotation of the rigid Earth with a multiplication factor of 0.76. The U.S.S.R. Time Service does not make such corrections. This is not a part of the reduction of observations because, in this case, the choice of the method of reduction defines the axis to which the ephemeris of the forced nutation refers. From our point of view the best choice of the reference axis would be the mean axis of figure of the Earth's mantle. A new set of nutation coefficients which would refer to this axis may be calculated on the basis of Molodenskij's theory of the rotation of the Earth. For example, the largest nutation coefficients adopted by the IAU and calculated by Molodenskij are given in Table 4. We have calcualted the amplitudes of the corresponding variations of time assuming the nutation coefficients given by Molodenskij are real and that $\tan \delta = 1.0$. These variations are shown in Table 5 for three cases:

I - the amplitudes of variations of time due to the errors of nutation terms and including the effect of forced diurnal motion of the pole computed for a rigid Earth,

II - the same as I but a deformable Earth was assumed for the derivation of the effects of forced diurnal motion,

III - the same as I without any corrections for forced diurnal motion.

Table 4. Values of nutation coefficients.

Argument	Adopted by IAU		Calculated by Molodenskij	
	$\Delta\epsilon$	$-\Delta\psi \sin\epsilon$	$\Delta\epsilon$	$-\Delta\psi \sin\epsilon$
Ω	9".210	6".858	9".203	6".841
2Θ	0.552	0.507	0.572	0.523
$2\mathcal{C}$	0.088	0.081	0.097	0.090

Table 5. Effect of the errors of nutation terms in time observations.

Argument	Aliasing Period (days)	Amplitudes of variations (ms)		
		I	II	III
$\Omega + \alpha$	385	0.33	0.33	0.33
$\Omega - \alpha$	346	0.73	0.74	0.80
$2\Theta - \alpha$	365	1.00	1.05	1.20
$2\Theta + \alpha$	122	0.13	0.13	0.13
$2\mathcal{C} - \alpha$	14.2	0.20	0.30	0.60
$2\mathcal{C} + \alpha$	13.2	0.00	0.00	0.00

4.2. Correction for the tidal variation of the vertical

Tidal variations in the direction of the vertical may be a cause of periodic variations in the observed longitude. These may be expressed in the form,

$$\Delta u_2 = \Lambda \frac{1}{g\rho \cos^2\phi} \frac{\partial V}{\partial \lambda}, \qquad (5)$$

where $\Lambda = 1 + k - \ell$ is the combination of Love and Shida numbers depending on the mechanical properties of the Earth, and ρ is the radius of the Earth, g is the acceleration of gravity, and V is the potential of the tide-producing force. The latter can be represented by the sum of periodic terms. These effects are shown in Table 6. The U.S.S.R. Time Service does not apply corrections for these effects. The BIH makes these corrections assuming that $\Lambda = 1.20$.

Table 6. Effects of the tidal variation of the vertical in time observations.

Tide	Aliasing Period (days)	Amplitude of Variations (ms)
K_1	365	0.73
P_1	183	0.24
K_1	183	0.16
M_1	14.8	1.26
O_1	14.2	0.52
N_1	10.3	0.24

4.3. Correction for errors in the polar coordinates

This correction is of the form

$$\Delta u_3 = (\Delta x \sin\lambda - \Delta y \cos\lambda) \tan\phi, \tag{6}$$

where Δx and Δy are the corrections to the polar coordinates. The motions of the CIO and the BIH 1968 reference systems relative to the MPO are caused mainly by the local effects at the stations participating in the ILS and the BIH. We have calculated the effects of these motions in time observations assuming that the MPO system is adopted as the standard. These are shown in Table 7.

Table 7. Effects of the instability of the reference systems in time observations (ms).

Component	CIO	BIH 1968
Linear trend	0.18/yr	--
22 to 30-yr. periodicity	2.40	--
10 to 12-yr. periodicity	1.20	--
Random	±0.60	±0.71

5. ON THE FREQUENCY RESPONSE OF THE FILTERS USED BY THE BIH AND THE U.S.S.R. TIME SERVICE

The systematic errors of time observations in the form (4) do not vanish in the determination of UT1 by the BIH and the U.S.S.R. Time Service because the observations are made approximately at the same local time and because the mean value of $\tan \phi$ for the participating observatories is 0.9 for the BIH and 1.3 for the U.S.S.R. Time Service. However, these errors are considerably reduced due to the following procedure for filtering individual observations:
  1) computation of the daily mean values,
  2) computation of the normal values, for example, 5-day mean values,
  3) averaging the results of all participating observatories,
  4) smoothing of the resultant determinations of UT1.

We can only find the approximate frequency responses of these filters because of the variety of programs of time observations and because of the variation in the weights of the normal values. Assuming the duration of observations during the day is five hours, the amplifying factors for the diurnal and semidiurnal variations will be 0.92 and 0.74 respectively. To estimate the amplifying factors of the other filters we have used the information given by Belotserkovskij and Kaufman (1979) and the BIH reports. The amplifying factors resulting from all of the filters mentioned above are given in Table 8. Use of Tables 5 through 8 allows us to estimate the different effects in UT1 which do not represent irregularities in the rotation of the Earth, but only defects in the processing and reduction of observations.

Table 8. Amplifying factors.

Component	Aliasing Period (days)	Series of Data			
		BIH		UT U.S.S.R.	
		unsmoothed	smoothed	unsmoothed	smoothed
Semi-diurnal	10.3	0.49	0.01	0.70	0.0
	14.8	0.61	0.05	0.74	0.0
	183	0.74	0.74	0.74	0.70
Diurnal	14.2	0.66	0.05	1.20	0.0
	122	0.82	0.82	1.20	1.02
	183	0.83	0.83	1.20	1.14
	346	0.83	0.83	1.20	1.20
	365	0.83	0.83	1.20	1.20
	385	0.83	0.83	1.20	1.20
Long-period	365 and longer	0.90	0.90	1.30	1.30

6. CONCLUSIONS

1. The existence of secular and long-period variations among the origins of the CIO, BIH and MPO systems does not support the reality of secular and long-period polar motion. These might be explained by local effects at the participating stations (in particular, the Ukiah VZT). For practical use in astronomy and geodesy the MPO is preferable.

2. The methods for preserving the reference systems by the BIH and the U.S.S.R. Time Service would be more effective if statistical predictions of the residuals for each station and instrument were applied.

3. Taking into account the frequency response to the filters used by the BIH and the U.S.S.R. Time Service, and since the shortest period in the tidal variations in the rate of the Earth's rotation is about 9 days it is advisable to compute UT1 for every day.

4. To reduce systematic errors in the determination of UT1 and to provide a comparison of results given by the BIH and the U.S.S.R. Time

Service it is necessary to apply a unified set of corrections for variations in the directions of the verticals.

REFERENCES

Belotserkovskij, D. Yu. and Kaufman, M. B.: 1979, this volume.
Fedorov, E. P.: 1975, Astrometrija i Astrofizika 27, pp. 3-6.
Fedorov, E. P., Korsun', A. A., Mironov, N. T.: 1972, in P. Melchior and S. Yumi (eds.), "IAU Symposium No. 48, Rotation of the Earth", D. Reidel, Dordrecht, pp. 78-85.
Markowitz, W.: 1970, in L. Mansihna, D. E. Smylie, A. E. Beck (eds.), "Earthquake Displacement Fields and the Rotation of the Earth", Springer-Verlag, New York, pp. 69-81.
Mikhailov, A. A.: 1971, Astron. Zu. 48, pp. 1301-1304.
Vicente, R. and Currie, R.: 1976, Geophys. J. Roy. Astron. Soc. 46, pp. 67-73.
Yumi, S. and Wako, Y.: 1966, Publ. Int. Latitude Obs. Mizusawa 5, pp. 61-86.

ETUDE D'OBSERVATIONS EFFECTUEES A L'ASTROLABE DE PARIS ET COMPARAISON
AVEC D'AUTRES RESULTATS CONCERNANT LES TERMES PRINCIPAUX DE LA NUTATION

Suzanne Débarbat
Observatoire de Paris

ABSTRACT

Chollet has analysed a homogeneous series of astrolabe observations
made at Paris since 1956.5 in order to derive corrections to the
principal terms of nutation. Comparison of his results with others
obtained recently indicates that the values $6\rlap{.}''840$ and $9\rlap{.}''210$ would be
suitable for adoption.

INTRODUCTION

Les observations examinées ici sont principalement celles obtenues
à l'astrolabe de l'Observatoire de Paris depuis 1956.5, date de l'installation d'un astrolabe de Danjon, remplacé en 1970.8 par un astrolabe
à pleine pupille. Le changement d'instrument n'a entraîné aucune modification du programme d'observations, dont les groupes sont demeurés
inchangés depuis cette époque.

Les données couvrent donc maintenant plus d'une révolution des
noeuds de la Lune et la réduction des observations présente le caractère d'homogénéité nécessaire aux études astrométriques. Ainsi le catalogue est le FK4, et le Système des constantes est celui imposé par
l'Union Astronomique Internationale.

Dès 1958, Guinot devait montrer (Guinot 1958) que le raccordement
des groupes d'étoiles observées à l'astrolabe de Danjon permettrait de
déterminer, avec une bonne précision, les constantes intervenant dans
la réduction des observations et, notamment, la constante de la nutation,
constante qui a donné lieu à des propositions de modifications à Kiev,
lors du Symposium UAI n° 78.

Les valeurs numériques n'ayant pas fait l'objet d'une décision
définitive, les valeurs qui sont données dans ce qui suit, constituent
des données supplémentaires à verser au dossier du mouvement des axes
auxquels sont rapportées les positions des étoiles lorsqu'on effectue

des déterminations de temps et de latitude à l'astrolabe.

LES OBSERVATIONS A L'ASTROLABE DE PARIS

La période couverte débute en juillet 1956 et comprend environ 6500 déterminations du temps et de la latitude. La précision des mesures avait été estimée (Guinot 1958) à $0^s.0043$ pour la première quantité, à $0".050$ pour la seconde, pour des groupes complets de 28 étoiles, après applications de corrections de lissage interne, ce qui est le cas des observations analysées (Chollet 1978).

Diverses corrélations ont été étudiées dans le passé (effet de magnitude, de type spectral, de marées terrestres...) faisant apparaître des effets relativement petits. O'Hora ayant fait remarquer (O'Hora 1977), qu'on pourrait craindre des effets liés aux conditions météorologiques, des observations de temps et de latitude, couvrant les six premières années d'observations à Paris, ont été étudiées dans ce sens. Les examens en fonction, respectivement, de la température et de la pression atmosphérique, de la direction et de la vitesse du vent, n'ont - pour le moment - montré aucune sensibilité particulière de ces mesures aux variables météorologiques.

Par ailleurs, on a examiné la stabilité à long terme des mesures effectuées à l'astrolabe de Paris et employées à la détermination des constantes de la nutation. Cette stabilité peut s'étudier par rapport à l'ensemble des instruments concourant aux travaux du Bureau International de l'Heure ou encore, de manière interne, selon la capacité de l'instrument à détecter les mouvements ou les irrégularités de la rotation de la Terre.

Du point de vue externe, on remarquera que le BIH avait attribué uniformément un poids de 100 en latitude et de 49 en temps à différentes stations équipées de PZT ou d'astrolabes, et ceci pour les années 1967 à 1970. A partir de 1971, le système de pondération change (Rapport annuel pour 1971, Annexe E) et dépend, pour une large part, non seulement de la qualité intrinsèque des observations, mais aussi de leur quantité.

Parmi les instruments qui conservent un poids confortable après 1970, on relève différents PZT dont celui en service à Herstmonceux dont on peut penser qu'il bénéficie de conditions climatiques voisines de celles de Paris (influence océanique). D'après les Rapports annuels du BIH, le poids moyen de l'une et l'autre station, calculé sur les données des six dernières années, se situe aux environ de 20 pour le temps. Pour la latitude, il est de 99 pour Herstmonceux et de 81 pour Paris. Ces valeurs placent ces stations dans un rang plus qu'honorable pour la stabilité à long terme recherchée par le BIH. L'astrolabe, toutefois, semble plus sensible à l'effet du manque d'observations ; il est vrai que la durée des groupes à l'astrolabe (une heure et demie) joue en sa défaveur en cas de conditions météorologiques médiocres.

La capacité de l'astrolabe à mettre en évidence les mouvements et les irrégularités de la rotation de la Terre constitue un critère de qualité interne de l'instrument. En ce qui concerne les mesures de latitude cette capacité est bien connue ; diverses analyses (Chollet et Débarbat 1972 et 1976, pour ne citer que les plus récentes) l'ont montré. Pour ce qui est du temps, elle a fait l'objet d'une autre étude (Chollet et Débarbat, à paraître) ; la conclusion est seule donnée ici : l'astrolabe peut détecter un changement dans la rotation de la Terre qui semble s'amorcer et ferait revenir à un régime sensiblement voisin de celui qui existait avant les modifications intervenues au début de 1973.

La stabilité des déterminations de temps et de latitude à l'astrolabe de Paris a permis de fournir déjà des valeurs des constantes de la nutation (Capitaine 1975 ; Capitaine 1977). Les résultats de latitude sont généralement considérés comme plus fiables ; cependant il apparaît que les résultats de temps présentent un caractère de stabilité qui les rend également propres à la détermination de constantes astronomiques, même si l'on a généralement tendance à penser que les mesures de latitude sont plus adéquates pour la détermination de certains paramètres intervenant dans le mouvement de la Terre.

Ce caractère de fiabilité des résultats, portant sur des analyses de mesures du temps et de la latitude effectuées à l'astrolabe de Paris, se trouve accentué par la prise en compte de variations de distance zénithale mises en évidence par Chollet sur approximativement la moitié des données (Chollet 1977), et appliquée à l'ensemble des mesures couvrant la période 1956.5 - 1978.0 (Chollet 1978).

COMPARAISON DES RESULTATS AVEC CEUX D'AUTRES STATIONS

Les résultats obtenus par différents auteurs et pour des déterminations récentes sont, d'une part, des analyses des mêmes données de base (astrolabe de Paris), d'autre part, des analyses portant sur le terme z tant des données du Service International des Latitudes que du Mouvement du Pôle, ou du Bureau International de l'Heure. Il existe aussi des analyses plus anciennes portant sur les données du Service International des Latitudes à partir des paires d'étoiles et pour lesquelles la durée des observations analysées est très longue.

Dans le Tableau I on a reporté, d'après Yokoyama et Chollet (Yokoyama 1977 ; Chollet 1978), dans les parties ABCD, les résultats obtenus selon l'origine de chaque analyse, ainsi que les valeurs UAI actuelles et les propositions de Kiev pour les termes principaux de la nutation (longitude, $N \sin \varepsilon$; obliquité O). Les erreurs ne sont pas mentionnées car elles sont généralement inférieures à la différence existant entre les diverses déterminations...

L'ensemble de ces résultats, déduits des observations astronomiques, présente des divergences qui atteignent 0".020, qu'il s'agisse de la nutation en longitude ou en obliquité, alors que les erreurs s'échelonnent de 0".002 à 0".011.

Pour l'astrolabe de Paris les valeurs obtenues à partir du temps sont généralement moins élevées que celles déduites de la latitude ; les valeurs moyennes qu'on en peut déduire sont, respectivement, 6".834 et 9".209 pour le temps, 6".838 et 9".213 pour la latitude.

CONCLUSION

Les valeurs numériques données ici pour les termes de nutation en longitude et en obliquité devraient permettre, avec la confrontation d'autres résultats, de concourir à la fixation des propositions définitives quant au choix des nouvelles constantes. Pour ce qui concerne les résultats déduits des observations à l'astrolabe de Paris, on doit remarquer que ce sont les seuls qui présentent le caractère d'homogénéité, aussi parfaite que possible, tant sur le plan des étoiles que du mode de réduction. On ne peut que regretter d'avoir couvert une seule révolution des noeuds de la Lune, et que d'autres instruments du même type et de précision équivalente ne puissent disposer d'une aussi longue série d'observations homogènes.

Pour ce qui est des valeurs proposées à Kiev, elles figurent, ainsi que les valeurs actuelles, dans la partie D du Tableau I. Il convient de remarquer que la valeur concernant la nutation en longitude est nettement au-dessus des valeurs de diverses autres analyses, qu'elles proviennent du terme z par le BIH ou l'ILS ou de toutes celles déduites de l'astrolabe de Paris. Pour la nutation en obliquité, la valeur proposée est nettement plus basse que celles déduites de l'analyse du terme z par le BIH et le IPMS, ainsi que l'astrolabe de Paris.

Pour ces deux termes nous proposerions que soit conservée la valeur UAI pour la seconde et que la première soit encore diminuée soit :
$$0 = 9".2100 \quad N \sin \mathcal{E} = 6".8400.$$

Ces valeurs correspondent à ce que peuvent fournir les déterminations astrométriques de ces termes, en l'état actuel des choses. Il conviendrait d'attendre les résultats déduits des nouvelles techniques d'observations pour tenter de gagner un facteur 10 dans la connaissance que nous avons de ces termes de nutation.

Les modèles de Terre, qui ne peuvent être que le reflet d'un ensemble de résultats d'observations, ne sauraient constituer que des indicateurs pour le choix de nouvelles valeurs. La valeur proposée pour la nutation en longitude serait d'ailleurs cohérente avec la plupart d'entre eux ; pour la nutation en obliquité, les modèles présentent presque tous une valeur à peine supérieure à 9".20, alors que les analyses récentes sont plus proches de 9".21. Il y aurait à rechercher les raisons de cette différence entre les valeurs proposées par les modèles théoriques et les valeurs déduites des observations astronomiques.

TABLEAU I

			N sin ε	O	Origine	
A	Yokoyama	1973	6".826	9".200	Terme Z	ILS
	Yokoyama	1977	6".842	9".209		IPMS
	Guinot-Feissel	1975	6".832	9".210		BIH
B	Fedorov	1963	6".8437	9".1980	Paires	ILS
	Taradia	1969	6".8476	9".1970	d'étoiles	ILS
C	Yokoyama	1975	6".838	9".212	Temps	Astrolabe
			6".831	9".212	Latitude	
	Capitaine	1977	6".831	9".205	Temps	de
			6".842	9".211	Latitude	
	Chollet	1978	6".833	9".209	Temps	Paris
			6".840	9".216	Latitude	
D	UAI		6".8584	9".2100		
	Symposium UAI n° 78		6".8430	9".2060		

Références

- Capitaine, N. 1975, Geophys. J.R. astr. soc., 43, p.573.
- Capitaine, N. 1977, Communication au Symposium UAI n° 78.
- Chollet, F. et Débarbat, S. 1972, Astron. and Astrophys., 18, p. 133.
- Chollet, F. et Débarbat, S. 1976, Wiss. Z. Techn. Univers. Dresden 25, H.4, p. 911.
- Chollet, F. 1977, Communication au Symposium UAI n° 78.
- Chollet, F. 1978, Communication au Symposium UAI n° 82.
- Guinot, G. 1958, Bull. Astron., XXII, p.1.
- O'Hora, N.P. 1977, Communication personnelle.
- Yokoyama, K. 1977, Communication au Symposium UAI n° 78.

A NEW SYSTEM OF THE U.S.S.R. STANDARD TIME FOR 1955 - 1974
AND ITS APPLICATION IN THE STUDY OF THE EARTH'S ROTATION

V. S. Gubanov and L. I. Yagudin
Pulkovo Observatory
Leningrad, U.S.S.R.

For the study of the Earth's rotation and geodetic measurements the longest possible series of homogeneous Universal Time (UT) data is required. For various reasons this remains a difficult problem. Periodically it is necessary to revise past observations and redetermine the UT scale. This was done by the Bureau International de l'Heure (BIH) in 1968 when the FK 4 catalog, revised values of astronomical longitudes, and new techniques were standardized for the reduction of astronomical observations. This system known as the 1968 BIH System has been in use since 1968 and has also been used for the re-reduction of observations back to 1962.0.

The U.S.S.R. Time Service, while participating in the BIH, also produces an independent UT scale, "Standard Time of U.S.S.R." (ST), for use in scientific and conventional geodetic work in the U.S.S.R. Since 1 January 1975 a new technique for the calculation of UT(ST) has been used. This involved the adjustment of longitudes as well as other measures to bring the new scale into agreement with UT(BIH) for the epoch 1975.0. Until now no attempt has been made to revise the data previous to 1975. These observations were essentially non-homogeneous due to past changes in origin, coordinate systems, and reference catalogs.

REVISION OF THE TIME SCALE

In the present work the data from 1955 through 1974 were used. During this period 26 observatories of the Soviet Union and other socialist countries participated in the U.S.S.R. United Time Service. The observations were made with 83 instruments including 40 visual instruments, 31 photoelectric transit instruments, eight astrolabes, three circumzenithals, and one photographic zenith tube. Approximately 78,000 clock corrections were determined by 300 observers. These initial data are the differences between the observed clock corrections and those adopted in the old system. The procedure followed in this revision was as follows:

1. All clock corrections were reduced to the KCB Catalog of Time Services (Pavlov, et al., 1971). Systematic as well as individual corrections were applied to the star coordinates.

2. The new value of the aberration constant (20".496) was used.

3. Observations made after 1962 were referred to the CIO and the BIH system. Data previous to 1962 were treated using the polar motion data of Fedorov and Korsun (1972) reduced to the BIH system by Brandt (Afanas'eva, et al., 1976).

4. A new method for deriving the new ST time scale from the data was used. Suppose that within the time interval, T, the time scale UT(ST) is defined by N adopted corrections to the reference clock, u_i, i = 1 to N. During this interval there exist M observational values, $u_{i,k}$, k = 1 to M. The observational values are in different systems of "instrument + observer" (IO). If we assume that during T the systematic errors of the IO systems are constant, then

$$u_{i,k} - \bar{u}_i = u_{j,k} - \bar{u}_j; \quad i,j = 1 \text{ to } N. \tag{1}$$

The values $\bar{u}_i$, $\bar{u}_j$ represent the true values of the corrections to the reference clocks for the new UT(ST) time scale corresponding to the times t_i and t_j.

Equation (1) can be written in the form

$$\Delta u_i - \Delta u_j = (u_{i,k} - u_i) - (u_{j,k} - u_j), \tag{2}$$

where $\Delta u_i = \bar{u}_i - u_i$, and $\Delta u_j = \bar{u}_j - u_j$. A system of $MN(N+1)/2$ equations of type (2) could be constructed for the interval T if the IO systems were not changed. This system of equations would have N unknowns Δu_i. The matrix of the resulting normal equations can not be inverted. Therefore it is necessary to adopt one value of the unknowns, for example, u_1, as an origin.

In reality the IO systems are constantly changing. The actual system of equations may then be constructed as follows:

4.1. Fortnightly means of the $u_{i,k}$ were formed for an interval of T = 2 years. The matrix of the resultant normal equations was of the order N = 48. The observations from 1955 - 1974 were then reduced by successive solutions of the matrices. Each successive solution overlapped the previous one by T/2 = 1 year.

4.2. The intervals, T_k, over which the IO system could be considered constant were found by studying the smoothed systematic errors, D_k. If there were no gaps in D_k during the interval T, and D_k was less than three times the mean square error of D_k, then T_k = T. Otherwise the interval was subdivided into segments of different duration.

4.3. Each equation (2) was assigned a weight to account for accidental errors and instability in D_k. The method is given by Yagudin (1978a).

4.4. The origin of the new UT(ST) scale was adjusted to that of the BIH by computing the systematic difference UT1(ST) - UT1(BIH) for the period 1962 - 1974.

The systematic differences between the old (ST_0) and the new ST scales and the BIH are shown in Figure 1. For comparison the new official scale of the U.S.S.R. Universal Time, UT1(SU) since 1968 using Kaufmann's

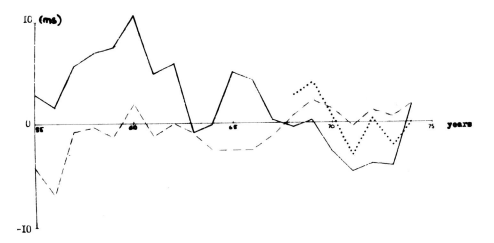

Figure 1. Systematic differences in UT1 time scales: ──────── UT1(ST_0) - UT1(BIH); ─ ─ ─ ─ UT1(ST) - UT1(BIH); ········ UT1(SU) - UT1(BIH).

method is also shown in Figure 1. Monthly differences in UT1(ST_0) - UT1(BIH) and UT1(ST) - UT1(BIH) were examined and found to be due mainly to the influence of different reference catalogs on the respective time scales.

CORRECTIONS OF LONGITUDES

The revision of these observations in a homogeneous system permits us to revise the longitudes of the contributing observatories. The corrections to the previously assumed longitudes were computed in relation to the UT(ST) time scale and the BIH scale. Both methods resulted in longitudes in excellent agreement (Yagudin, 1978b).

LONG-PERIOD VARIATIONS IN THE RATE OF ROTATION

Variations of the rate of rotation relative to the BIH atomic time scale were calculated using unsmoothed fortnightly values of UT1(ST) in the form $\tau_i = (UT1 - AT)_i$, and

$$\left(\frac{d\tau}{dt}\right)_i = \frac{\tau_{i+1} - \tau_{i-1}}{AT_{i+1} - AT_{i-1}}.$$

The data from 1956 to 1974 were divided into three seven-year periods, and the spectra of $d\tau/dt$ were treated separately to obtain independent evaluations. The significant periodic terms are shown in Table 1. These results are determined from the entire data set from 1956 to 1974. The seasonal variation (UT2 - UT1) was found to be

$$\Delta T_s = 0\overset{s}{.}016 \sin 2\pi t - 0\overset{s}{.}012 \cos 2\pi t - 0\overset{s}{.}003 \sin 4\pi t + 0\overset{s}{.}008 \cos 4\pi t,$$

where t is the fraction of the Besselian year.

Table 1. Variation in $d\tau/dt$ and τ in the form $A \sin (2\pi/P + \phi)$ using UT1(ST) from 1956 to 1974. Errors are 90% probable errors.

	$d\tau/\tau dt$		τ	
P (years)	A (10^{-9})	(degrees)	A ($0\overset{s}{.}001$)	(degrees)
1.00 ± 0.01	4.0 ± 0.3	59 ± 4	20 ± 2	149 ± 4
0.50 ± 0.01	3.4 ± 0.3	213 ± 5	8 ± 1	303 ± 5
2.9 ± 0.2	0.9 ± 0.3	236 ± 19	13 + 4	329 ± 19
3.7 + 0.4 − 0.3	0.6 ± 0.3	244 ± 26	11 ± 6	334 ± 26
2.0 ± 0.1	0.5 ± 0.3	87 ± 31	5 ± 3	177 ± 31
6.9 + 1.0 − 0.8	2.3 ± 0.3	316 ± 7	80 ± 10	46 ± 7
11 + 5 − 3	1.4 ± 0.3	54 ± 12	77 ± 17	144 ± 12

SHORT-PERIOD VARIATIONS IN THE RATE OF ROTATION

Residuals of the form $\Delta u_{t,k} = u_{t,k} - \bar{u}_t - D_{t,k}$ were analyzed for short-period variations. Values of u_t and $D_{t,k}$ were derived by smoothing normal values for every 15 days using Whittaker's method (Yusupov) with a smoothing factor $\varepsilon = 0.1$. Amplitude spectra computed with the Fast Fourier Transform show periods which can be identified with the M_m, M_f, O_1, and M_2 tides. The details are given by Gubanov and Yagudin (1978). Values of the Love number k derived from these results are:

$k(M_m) = 0.31 \pm 0.02$, and
$k(M_f) = 0.30 \pm 0.02$.

These are in good agreement with the predictions of Molodenskij Earth Model II (Molodenskij, 1961). These results can also be used to derive estimates for the combination of Love and Shida numbers $\Lambda = 1+k-1$. For the M_2 wave $\Lambda = 1.10 \pm 0.05$, and for the O_1 wave $\Lambda = 1.39 \pm 0.10$. The latter result was derived using the theoretical value of the forced luni-solar polar motion. If a value for Λ is assumed to be 1.20 an improved value for the fortnightly nutation in obliquity is found to be $0.''0897 \pm 0.''0007$.

REFERENCES

Afanas'eva, P. M., Brandt, V. Eh., Efremova, N. P., Pavlov, N. H.: 1976, Pis'ma v Astron. Zh. 2, pp. 452-454.
Fedorov, E. P., Korsun, A. A.: 1972, "Dvizhenie Polyusa Ehemli s 1890.0 po 1969.0", Kiev.
Gubanov, V. S., Yagudin, L. I.: 1978, Pis'ma v Astron. Zh. 4, p. 108.
Molodenskij, M. S., Kramer, M. V.: 1961, "Ehemnye Privily i Nutatsiya Ehemli", Iehd-vo Acad. Nauk U.S.S.R., M.
Pavlov, N. N., Afanas'eva, P. M., Staritsyn, G. V.: 1971, Trudy Gl. Astron. Obs. Pulkove ser. II, vol. 78, pp. 59-98.
Yagudin, L. I.: 1978a, "Vyiod Novoj Shkaly Vsemirnogo Vremini eha 1955-1974", Astrometriya i Astrofienika (Resp. Mezhved. Sb.), no. 35.
Yagudin, L. I.: 1978b, "Popravki Dolgot Sluzhb Vremeni U.S.S.R.", Astrometriya i Astrofienika (Resp. Mezhved. Sb.), no. 35.
Yusupov, Yu. G.: Iehvestiya Astron. Ehigel'gartovsk Obs., no. 36, p. 229.

DISCUSSION

S. Debarbat: Was it your intention to produce a time scale as close as possible to that of the BIH?
V. S. Gubanov: Our scale has been adjusted to that of the BIH only in the average over the entire time interval.

AN ANALYSIS OF THE ROTATIONAL ACCELERATION OF THE EARTH

Dennis D. McCarthy and Donald B. Percival
U. S. Naval Observatory
Washington, D. C. 20390 USA

ABSTRACT

The nature of the irregular fluctuations in the speed of the Earth's rotation was investigated using ninety-day means of UT2 - A.1 determined at the U. S. Naval Observatory. Data from June 1955 to April 1978 were included in this study. Statistical analysis of the excess length of day shows no evidence for the persistence of discrete values for periods on the order of five years. No statistical basis for the existence of discrete "turning points" in the rotational speed could be found.

Spectral analysis of the acceleration data shows that the rotational acceleration of the Earth during this period of time may be represented by a constant term plus random changes in acceleration occurring with a frequency greater than once per year. The magnitude of these changes appear to be consistent with estimates of meteorologically induced changes in the rotational acceleration.

An autoregressive integrated moving average (ARIMA) model was fit to the excess length of day data. This model permits simulated series of excess length of day data to be constructed. These simulated series show a statistical similarity to observations made since 1820. However the apparently large changes in the acceleration which occurred around 1870 and 1900 are twice that which can be reasonably accounted for by this model. The details of this analysis will be published later.

DISCUSSION

C. R. Wilson: Why did you smoothe the data at all? How did you decide upon the order of the ARIMA model? How did you treat the early data which were given at a different sample interval?
D. D. McCarthy: The daily values were smoothed to eliminate what was thought to be short-period observational noise. In view of these results perhaps less smoothing would be desirable.

The order of the ARIMA model used for the estimation of the spectrum and for the simulation was determined from a general information theory criterion due to Akaike and explicity formulated by Ozaki. For prediction purposes, a much simpler model was found to be sufficient. This model was selected using the techniques described by Box and Jenkins.

For the purpose of this paper, the early data were not modeled explicitly. Instead, we took the ARIMA model determined from the last 23 years of observed data, created simulated sequences extending for 150 years, and looked at the peak-to-peak dispersions in the simulated series as compared to the historical data. From this work, we concluded that the ARIMA model could reasonably explain only about half of the peak-to-peak dispersion actually seen in the early historical data.

S. Debarbat: Chollet and myself have begun a similar analysis with the Paris astrolabe observations. I only mentioned this fact in my paper, but it will be very interesting to compare our results with yours.

S. K. Runcorn: Would you like to comment on the possible geophysical interpretation of your data on the length-of-day variations as an accumulation of random impulses?

D. D. McCarthy: Our analysis shows that the length-of-day variations can be simulated with random accelerations occurring more frequently than once per year. We have not attempted to relate these to a geophysical cause. However, the accelerations required appear to be consistent with those which can be produced meteorologically.

NEW DETERMINATION OF THE "DECADE" FLUCTUATIONS IN THE ROTATION OF THE EARTH 1860-1978

L. V. Morrison
Royal Greenwich Observatory, U.K.

Abstract[*]

Observations of the Earth's rotation have shown irregular variations of rate which have characteristic times of decades. These have been attributed to transfer of angular momentum between core and mantle by some mechanism such as inertial coupling, viscous stress, electromagnetic coupling or stresses produced by topographic features on the core mantle boundary.

Possible mechanisms must be tested against observed fluctuations in the Earth's rotation, such as those derived by Brouwer in 1952. His results were based on the observed departure of the Moon from its predicted position relative to the stars, and it occurred to us that it ought to be possible to improve upon the accuracy of Brouwer's results by analysing more observations than were available to him and by correcting occultation data for the irregularities of the Moon's outline.

In conjunction with the US Naval Observatory we have collected and analysed 50 000 independent observations of occultations of stars by the Moon in the years 1861 to 1954, and our results for the variations in the rotation of the Earth are presented here. I hope that they will provide geophysicists with tighter and more reliable constraints on their models for core-mantle coupling.

Each occultation observation provides a measure of the cumulative rotational displacement $\Delta\theta$ of the Earth from a fixed direction. In order to simplify the subsequent analysis, observations within individual years have been combined to obtain independent annual solutions for the angle $\Delta\theta$. For 1955 and later years we have derived similar annual solutions from the BIH values for the difference between UT and atomic time. Our points are less erratic than Brouwer's and they tend to lie on a smooth curve.

[*]The full paper will be published in Geophys. J.R. astr. Soc., (1979).

First and second time derivatives of $\Delta\theta$ have been estimated from these annual solutions to obtain the length-of-day and its rate of change throughout the entire interval. The derivatives at a given date were obtained from least squares fits of quadratic functions to, respectively, 5 and 11 consecutive points centered on that date.

The magnitude of the torque on the mantle is given by the product of the observed angular acceleration and the principal moment of inertia. Around 1900 the torque changed by 10^{18} newton metres in about 5 years. The torque curves are similar in the periods between 1930-1950 and 1955-1975: the time reference during the former period was based on the lunar motion, whereas atomic time was used throughout the second period.

Estimation of the time derivatives in this work has necessarily entailed some smoothing of the data, and there was the risk that significant detail might be lost. The 11-point estimator was chosen in the light of a power spectrum analysis of the second differences of the annual values of $\Delta\theta$; this showed a broad peak in power around a period of 30 years, and a general rise in power at shorter periods which was to be expected in a spectrum formed from second differences. There are no significant peaks of power at periods shorter than about 30 years, except for the rise at the Nyquist period of 2 years which I interpret as evidence of the annual variation.

In order to check this interpretation of the occultation results I have performed a power spectrum analysis of the values of $\Delta\theta$ taken at 50 day intervals from the BIH results for TAI-UT2. The spectrum shows significant peaks of power at periods of $\frac{1}{2}$ year and 1 year, caused by random fluctuations in the amplitude of the seasonal variation, but there are no significant peaks at periods between 1 and 8 years. The time-range of the data is too short to draw any conclusions about periods longer than 8 years, but the results are not incompatible with my conclusion from the occultation data that most of the power in the decade fluctuations is at periods longer than 30 years.

DISCUSSION:

S.K. Runcorn: Would you comment on the possible significance of the shortness of the interval in which changes of slope of the $\Delta\theta$ curve become established, especially just before 1900?

L.V. Morrison: The power spectrum analysis of the independent annual values of $\Delta\theta$ shows no significant peaks in the power at periods shorter than about 30 years.

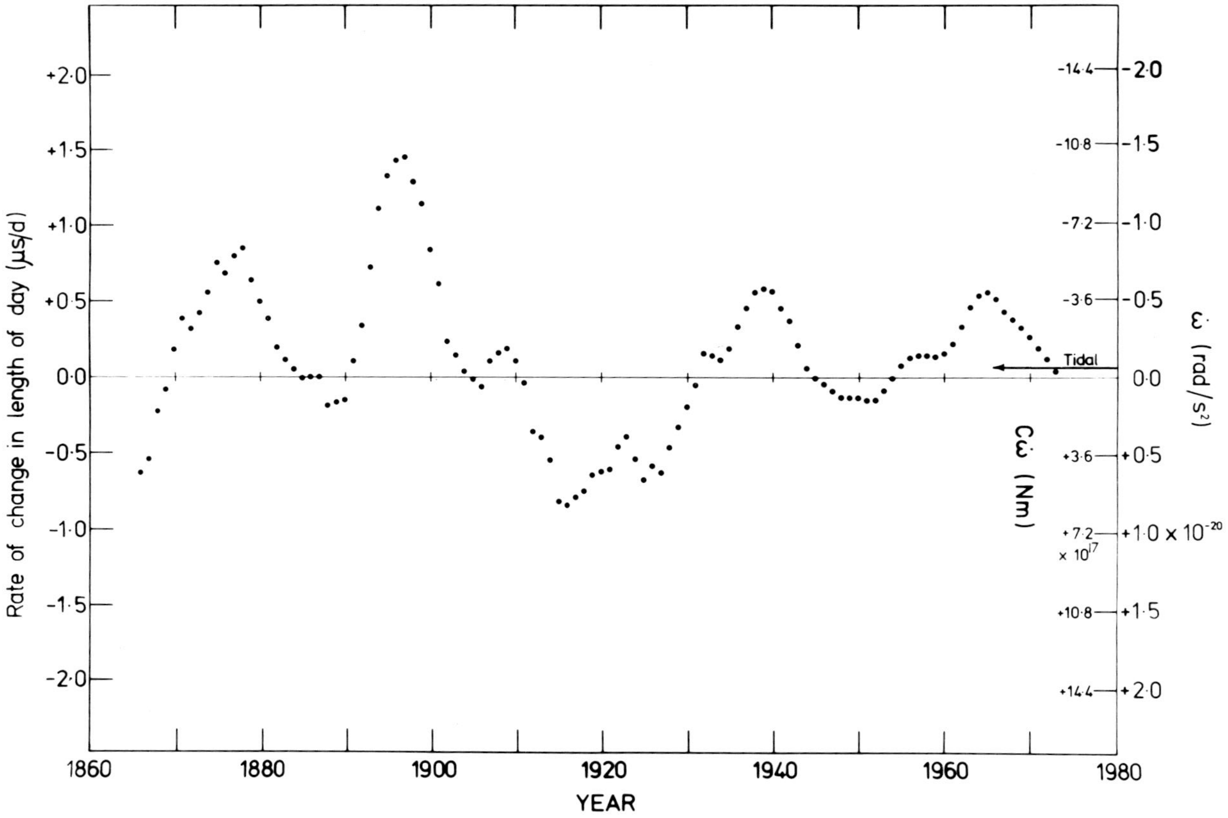

"DECADE" FLUCTUATIONS IN THE EARTH'S ROTATION 1860–1978   57

Figure 1. The second time derivative of $\Delta\Theta$, obtained by fitting a moving quadratic polynomial to 11 successive annual values. The inside right-hand ordinate measures the torque operating on the mantle in units of 10^{17} newton metres. The arrow marked "tidal" indicates the level of the combined lunar, solar and atmospheric tidal torque.

ON THE LONG-PERIOD VARIATIONS IN THE RATE OF THE EARTH'S ROTATION

A. I. Emetz and A. A. Korsun'
Main Astronomical Observatory
Ukrainian Academy of Sciences
Kiev, U.S.S.R.

ABSTRACT

The maximum entropy power spectrum (Smylie, et al., 1974) of the Earth's rotational speed was calculated using data from 1900 to 1976. Two series of data were analyzed. The first was a series of $\delta\omega/\omega$ determined from annual UT1 - ET data from 1900 to 1976. The second was a similar series derived from the mean monthly data of UT1 - TAI. Linear trends were removed from both series before analysis. Using the second series of data, significant periods of 2.8, 3.7, 7.0, and 10.5 years were found. The first series showed significant periods at 6, 10, 13, 22, and 57 years. Of these periodicities those at 22 and 57 years showed the largest amplitudes (0.454 ± 0.097 x 10^{-8} and 1.431 ± 0.104 x 10^{-8} respectively).

Various geophysical phenomena such as global motions in the oceans (Naito and Kikuchi, 1973), topographic coupling at the core-mantle boundary (Smith, 1974), variation in the intensity of zonal circulation of the atmosphere (Lambeck and Cazenave, 1974), have been proposed to account for variations in the Earth's rotation with periods from 2 to 10 years. The effect of solar activity may also be a cause of the long-period variations in the Earth's rotation. Afanas'eva, et al. (1965) have shown that the solar wind can interact with the Earth's magnetosphere to induce current within it. A small portion of corpuscular stream energy is sufficient to cause changes in the speed of the Earth's rotation. Afanas'eva (1966a, b) has found a correlation between the 22-year cycle of solar activity and similar variations in the Earth's rotation. Magnetohydrodynamic twisting oscillations in the core of the Earth may account for variations in rotational speed with periods of about 60 years (Braginsky, 1970). While Jady (1969) has shown that solar wind energy is insufficient to explain variations in the Earth's rotation, Kalinin and Kiselev (1977) showed that solar activity may be a cause of such oscillations and that there is a correlation between a 60-year variation in the Wolf numbers and a 60-year variation of $\delta\omega/\omega$.

The nature of the variation of the Earth's rotational speed occurring with periods of 22 and 60 years needs further study. It is particularly important to construct a quantative theory of the interaction between the solar wind and magnetosphere and, through it, with the mantle which can account for the effect of the interplanetary field. More complete information on the space and time variations of the solar wind are also required.

REFERENCES

Afanas'eva, P. M., Kalinin, U. D., Molodenskij, M. S.: 1965, Geomagnetizm i Aeronomija 5, p. 795.
Afanas'eva, P. M.: 1966a, Geomagnetizm i Aeronomija 6, pp. 611-613.
Afanas'eva, P. M.: 1966b, Geomagnetizm i Aeronomija 6, p. 944.
Braginsky, S. I.: 1970, Geomagnetizm i Aeronomija 10, pp. 3-7.
Kalinin, Ya. D., Kiselev, V. M.: 1977, Geomagnetizm i Aeronomija 17, pp. 166-167.
Jady, R. J.: 1969, in L. Mansinha, D. E. Smylie, and A. E. Beck (eds.), "Earthquake Displacement Fields and the Rotation of the Earth", Springer-Verlag, New York, pp. 115-121.
Lambeck, K., Cazenave, A.: 1974, Geophys. J. Roy. Astron. Soc. 38, pp. 49-61.
Naito, I., Kikuchi, N.: 1973, Proc. Int. Latitude Obs. Mizusawa 13, pp. 179-191.
Smith, J.: 1974, VDI Nachr. 28, p. 31.
Smylie, D. E., Clarke, G. K., Ulrych, T.: 1973, Methods of Computational Physics, vol. 13, pp. 391-430.

DISCUSSION

L. V. Morrison: I wonder how significant the period found at 10 years is? I did not find this in the spectral analysis of my new data.
Ya. S. Yatskiv: The most significant periods are 22 and 57 years.
R. O. Vicente: Analysis of the ILS data extending over 75 years by a maximum entropy method shows a period of about 30 years.

MAIN RESULTS OF STUDYING THE NATURE OF THE IRREGULARITY OF
THE EARTH'S ROTATION

N. S. Sidorenkov
U.S.S.R. Meteorological Center

The variations of the atmospheric angular momentum were investigated (Sidorenkov, 1976). Using the climatic cross-sections of the zonal wind, the values of the relative angular momentum of the atmosphere, h, were calculated for each month. The variations of h during the year are shown in Figure 1, where curve 1 illustrates the sum of h for the entire atmosphere, and curves 2 and 3 illustrate h for the atmospheres of the northern and southern hemispheres respectively.

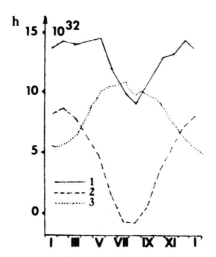

Figure 1. Annual variations in the relative angular momentum of the atmosphere: 1 - the entire atmosphere; 2 - the northern hemisphere; 3 - the southern hemisphere.

From the data on the relative atmospheric angular momentum we can easily derive the corresponding variations in the rate of the Earth's rotation, $\delta \omega / \omega = - \delta h / I \omega$, where ω is the augular velocity of the Earth's rotation, and I is the Earth's moment of inertia. Figure 2 shows both the calculated (curve 3) and the observed seasonal variations, $\delta\omega/\omega$ averaged for the period 1956-1961 (curve 2), and for the period 1962-1972 (curve 1). Comparison of the curves confirms the hypothesis that the seasonal irregularity of the Earth's rotation is due to the seasonal variation of h.

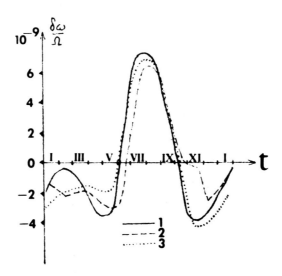

Figure 2. Annual variations in the rate of the Earth's rotation obtained from the data of the atmospheric angular momentum (curve 3), and the astronomical observations averaged for the period 1962-1972 (curve 1), and for 1956-1961 (curve 2).

The correlation of h with the seasonal irregularity of the Earth's rotation yields essential information concerning the atmospheric processes. Studying the nature of seasonal variations, the author discovered a previously unknown thermal engine in the atmosphere - the "inter-hemisphere engine" (Sidorenkov, 1975). This is caused by different seasonal atmospheric heating in the northern and southern hemispheres. As thermodynamic analysis shows, this engine results in the seasonal variation of h and consequently the seasonal irregularity of the Earth's rotation. These seasonal variations are not the sum of annual and semiannual harmonics as had been considered before, but can be written in the form,

$$\frac{\delta\omega}{\omega} = -\frac{\delta h}{I\omega} \sim |\Pi + E \cos(\Theta - \beta)|,$$

where Π and E are constants, Θ is the "longitude" of the Sun reckoned from the beginning of the year, and β is the phase. For example, the average for the period 1962-1972 of the seasonal irregularity of the Earth's rotation is given as

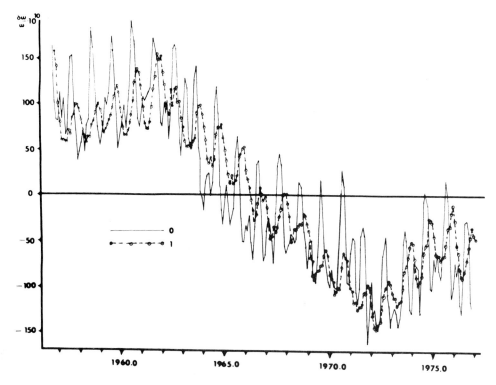

Figure 3. Variations of the mean monthly values of the rate of the Earth's rotation during the recent 20 years: 0 - from the astronomical data; 1 - from theoretical values.

$$\frac{\delta \omega}{\omega} \times 10^{10} = |30 + 89 \cos(\Theta - 201°)|.$$

The inter-hemisphere thermal engine reduces the dominant atmospheric air transport from West to East. Twice a year, in April and November, when the temperature in the northern and southern hemispheres is nearly the same, the effect of the inter-hemisphere engine is absent. Then h is maximum, and the rate of the Earth's rotation consequently is reduced to a minimum. In July and January the opposite effect is observed. Because the temperature difference is more in July than in January h is decreased, and the corresponding increase of the Earth's rotation is greater in July than in January.

The results of the study of the mechanical effect of the atmosphere on the Earth are given by Sidorenkov (1978). The monthly values of atmospheric pressure at ground level for the period 1956.8 - 1977.0 were calculated. These data enabled us to derive the variations of the Earth's rotation due to the friction of the atmosphere on the land surface and its pressure upon mountains. The theoretical curve of the variation of the Earth's rotation as well as the observed irregularity of the Earth's rotation are given in Figure 3. Both curves are rather similar with regard to yearly variations. The seasonal variations cannot be obtained in such a manner because the initial assumption in this study was not sufficiently correct.

The above calculations show that the yearly variations of the Earth's rotation are caused by a mechanical atmospheric effect on the Earth. This conclusion causes a problem regarding the balance of the angular momentum of the Earth. In our opinion the Earth-atmosphere system is not closed. It is possible that the flow of small portions of positive or negative angular momentum from space to the atmosphere exists. They cannot be retained in the atmosphere and flow down to the Earth through the near-surface layer. These portions are accumulated by the Earth, and may cause yearly variations in the rate of the Earth's rotation.

REFERENCES

Sidorenkov, N. S.: 1977, Izvestija Acad. Nauk U.S.S.R. Fizika Atmosfery i Okeana 12, pp. 351-356.
Sidorenkov, N. S.: 1975, Dokl. Akad. Nauk U.S.S.R. 221, p. 4.
Sidorenkov, N. S.: 1978, Astron. Zu. 55.

DISCUSSION

C. R. Wilson: Does this theory imply that the thermodynamic engine causes secular variation of the rotation?
Ya. S. Yatskiv: No. It was not investigated.

SOME ADVANTAGES AND DISADVANTAGES OF A PHOTOGRAPHIC ZENITH TUBE

Dennis D. McCarthy
U. S. Naval Observatory
Washington, D. C. 20390 USA

ABSTRACT

The theory and instrumentation of a photographic zenith tube (PZT) have been described by various authors (Torao, 1959; Markowitz, 1960; Thomas, 1964; Schuler, 1968; Takagi, 1974). Analysis of the more recent results of the PZT's of the U. S. Naval Observatory at Washington, D. C. and Richmond, Florida, permit the determination of precision and accuracy estimates which may be expected from the daily operation of a PZT. This analysis shows precisions of $\pm 0\rlap{.}''03$ in latitude and $\pm 0\rlap{.}^s003$ in time, determined from the internal error of one sight. Accuracies estimated from the external error of a sight were found to be $\pm 0\rlap{.}''08$ in latitude and $\pm 0\rlap{.}^s006$ in time.

In comparison to other techniques available for the measurement of Earth rotation parameters, the PZT has many advantages. Among these are its rigid construction, elimination of levelling errors, minimization of refraction errors, low cost, and ease of operation. This instrument also observes stars whose positions may be related to a visual fundamental reference frame. Conversely the observations may be used for improvement of this reference frame. The use of PZT's in established chains eliminates the need for the development of more star catalogs and minimizes the effect of systematic catalog errors in the determination of polar motion.

Among the disadvantages of a PZT are its limited field of view and possible systematic errors in the positions and proper motions of the internal star catalog. Also the observations are limited to clear skies and by systematic atmospheric refraction effects.

These advantages and disadvantages along with the well-established estimates of the instrument's accuracy must be considered in the development of future instrumentation for the study of the rotation of the Earth.

REFERENCES

Markowitz, Wm.: 1960, "Stars and Stellar Systems", vol. 1, p. 88.
Schuler, W.: 1977, "Etude Theorique et Experimentale de la Lunette Zenithale Photographique (PZT) de Neuchatel", Edition Medicine & Hygiene, Geneva.
Takagi, S.: 1974, Pub. Int. Latitude Obs. of Mizusawa 9, p. 259.
Thomas, D. V.: 1964, Royal Obs. Bulletins No. 81.
Torao, M.: 1959, Ann. Tokyo Astron. Obs. Second Series 6, p. 103.

DISCUSSION

S. Debarbat: You have mentioned the 65-cm PZT of the U. S. Naval Observatory in your talk; can you say something about it?
D. D. McCarthy: Analysis of the preliminary observational results of the 65-cm PZT indicates that at least a 25% improvement in precision may be expected. We are currently working on the elimination of a systematic temperature effect in the rotation angle and the improvement of the environmental control. The original instrumental design has been changed to incorporate improvements in the plate drive mechanism and computer control of the instrument. When the 65-cm PZT becomes fully operational (hopefully in the latter part of 1978) it will be used to observe all of the stars which have been observed by the Washington and Mizusawa PZTs in the past along with a supplemental list of stars numbering approximately one thousand.

TIDAL PERTURBATIONS IN ASTRONOMICAL OBSERVATIONS

N.P.J. O'Hora
Royal Greenwich Observatory, Herstmonceux

T.F. Baker
Institute of Oceanographic Sciences, Bidston Observatory

SUMMARY. Twenty years of time and latitude observations made with the Herstmonceux Photographic Zenith Tube have been analysed to determine the phase and amplitude of deflexions of the local vertical at the frequency of the semi-diurnal M_2 tide. The results have been compared with predictions incorporating new calculations of the gravitational attraction due to the M_2 tides in the surrounding seas. The effects of the zonal M_m and M_f tides on the time observations have also been determined and compared with existing theory. In both cases the agreement is reasonably good.

Group results of the time and latitude observations made with the Herstmonceux Photographic Zenith Tube (PZT) have been analysed to determine their tidal components. The data analysed extend over the years 1958-1977; they are based on observations of 36000 star transits giving 5800 group results. The time observations are expressed as measures of UT0 - TA (RGO), where TA (RGO) is the atomic time scale of the Royal Greenwich Observatory. The latitude observations have been corrected for the diurnal drift in latitude at Herstmonceux (Thomas and Wallis, 1971). The two sets of observations were smoothed by the Vondrák method (Vondrák, 1969) to obtain smooth values that are almost unaffected by periodic terms of less than 30 days period. Observational residuals, in the sense "observed minus smoothed" were then calculated.

Tidal deflexions of the vertical that affect PZT observations are primarily due to the global body tide of the Earth and are the direct effect of variations in the tidal force as the direction of the Moon changes. There is also an oceanic effect due to variations in the horizontal gravitational attraction, caused by the tidal variation in sea level. In the case of the Herstmonceux PZT, the changes in the direction of the vertical caused by the oceanic effect are of the same order of magnitude as those due to the direct effect. For the principal lunar semi-diurnal component M_2, the deflexions are periodic functions of 2H, where H is the hour angle of the Moon and, in the case of the direct effect, they are proportional to Λ, where $\Lambda = (1+\kappa-\ell)$ and κ and ℓ are body tide Love numbers.

The amplitudes of deflexions affecting the PZT observations in the East-West and in the North-South directions were determined by least-squares analyses of the time and latitude residuals of the group results respectively. Earlier determinations of the deflexions (O'Hora, 1973; O'Hora and Griffin, 1977) were obtained by analyses of plate residuals. But since a plate residual represents an average value based upon several hours of observation, the effects in a plate result of semi-diurnal term are diminished by smoothing. Group results represent, on average, slightly more than an hour's observing time so they conserve the effects of such a term virtually free from smoothing. The deflexions of the downward vertical given by the new analyses of the time and latitude residuals are, in units of milliseconds of arc, respectively:

$$16.1 \cos (2H - 118°) \text{ towards the West}$$
$$\text{and } 22.0 \cos (2H + 31°) \text{ towards the South}$$

with standard errors of 1.0 in amplitude and $4°$ in phase.

The deflexions due to the oceanic effect in these directions were evaluated by dividing the surrounding seas into sections and calculating for each the amplitude and phase of the horizontal component of the gravitational attraction at Herstmonceux due to the M_2 ocean tide. Figure 1 shows the M_2 tidal attractions at Herstmonceux from the different sea areas. The largest North-South contribution arises from the eastern English Channel only 7 kilometres distant; significant East-West contributions arise from the southern North Sea, the English Channel, the Celtic Sea and the North Atlantic Ocean. Figure 2 shows M_2 phasor plots of the observations and the theoretical calculations for the west and south directions respectively. The amplitudes of the deflexions in milliseconds of arc are plotted radially and the phase lags (α) with respect to the argument of the direct tidal potential at Herstmonceux are plotted anti-clockwise. The uncertainty in the theoretical global body tide due to the range of possible seismic Earth models is less than $\pm 1\%$ (see, for example, Farrell 1972). For these calculations a value of $\Lambda = 1.22$ has been used. In the plots, the total theoretical deflexion (body tide + sum of attractions of the different sea areas) is compared with the observed values. The observed error circles are standard errors calculated from the least-squares solutions.

The importance of the oceanic tidal attractions is very clear and, in both directions, their inclusion considerably improves the agreement between theory and observation. The main uncertainty in the theoretical calculations arises from the uncertainty in the amplitudes and phases of the M_2 ocean tides. These are of the order of $\pm 10\%$ in amplitude and $\pm 10°$ in phase (Baker 1977). The effects on the PZT of the tidal deformation of the Earth due to ocean-tide loading have been neglected.

TIDAL PERTURBATIONS IN ASTRONOMICAL OBSERVATIONS   69

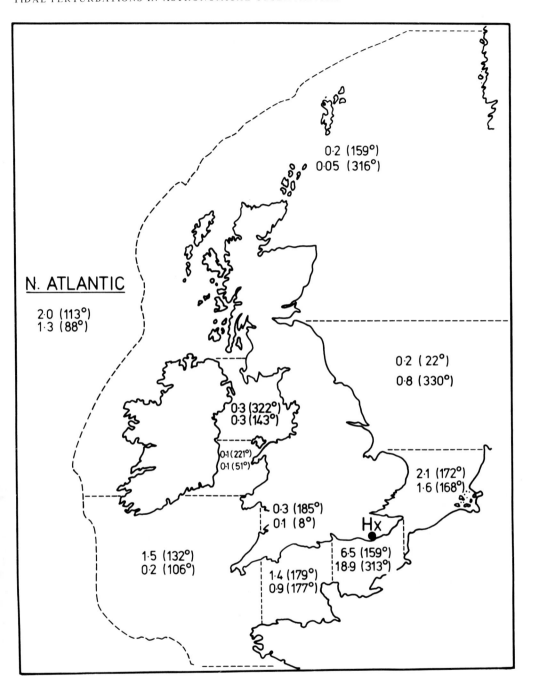

Figure 1: Deflexions of the downward vertical at Herstmonceux (Hx) due to the M_2 tide in different sea areas. Amplitudes are in milliseconds of arc and phase lags are with respect to 2H, the argument of the tidal potential at Herstmonceux. Upper lines are west deflexions, lower lines are south deflexions.

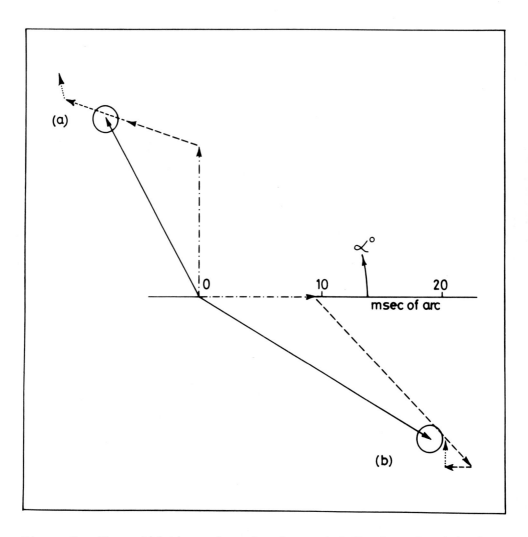

Figure 2: The solid lines show the observed deflexions for (a) time and (b) latitude, while the broken lines show calculated deflexions as follows:- (a) West deflexion; tilts are positive when a mass on a plumb line moves towards the west. (b) South deflexion; tilts are positive when the plumb line moves towards the south. In both cases the deflexion is in msec of arc and α is the phase lag with respect to the argument of the direct tidal potential at Herstmonceux.

—·—·—·—·—	theoretical global body tide, $\Lambda = 1.22$
— — — — —	east English Channel attraction
- - - - - - -	rest of continental shelf attraction
·················	N. Atlantic attraction

TIDAL PERTURBATIONS IN ASTRONOMICAL OBSERVATIONS

Firstly, the change in potential of the deformed Earth gives rise to a horizontal attraction described by the load Love numbers k_n (Farrell 1972). Secondly, the horizontal displacement of the PZT gives a deflexion of the vertical described by the load Love numbers ℓ_n. These effects of the loading deformation are small compared to the gravitational attraction of the ocean tide which produces them. Order of magnitude estimates based on the load Love numbers of Farrell show that the total oceanic effects in Figures 2 (a) and (b) should be reduced by approximately 10% and 5%, respectively. Allowing for the above uncertainties in the oceanic effects there is very good agreement between theory and observation, particularly in the North-South direction.

The zonal tides induced by the Moon cause the moment of inertia and, consequently, the speed of rotation of the Earth to vary, and the variations are reflected in the time observations. These tides do not influence the position of the pole of inertia so the latitude observations are not affected. Melchior has recently tabulated theoretical values for the terms in the time observations (Melchior, 1978). The amplitudes of these terms are all proportional to k, and assuming $k = 0.30$, the amplitudes of the principal terms are 0.75, 0.31 and 0.80 ms with periods, in days, of 13.661 (M_f term) 13.633 and 27.555 (M_m term), respectively. The PZT time residuals were analysed to determine the amplitudes of the periodic terms corresponding to the M_f and M_m tides. Before the least-squares analysis the residuals were corrected for the semi-diurnal deflexions found above, and also for the effects of a 14.19-day term, with amplitude 1.27 ms, due to nutation and aliasing of the O_1 diurnal tide. The amplitudes obtained for the M_f and M_m terms are 1.06 ± 0.12 and 0.84 ± 0.12 ms respectively. The disparity in the results was investigated analysing the first and third quarters of the data separately from the second and fourth quarters.

These solutions yielded amplitudes of 1.12 and 0.93 for M_f and 0.58 and 1.15 ms for M_m, all with standard errors of about 0.17. For each term the phases given by the half-data solutions are in much better agreement than the amplitudes. The influence of excessive residuals in the data was examined by rejecting all residuals outside the range $0 \pm 40n^{-\frac{1}{2}}$ ms where n is the number of stars in the group observation and 40 ms is approximately 4 times the standard error of unit weight. Analyses of the remaining 99% of the data gave amplitudes of 0.98 and 0.85 for M_f and 0.55 and 1.03 ms for M_m. Adopting the means of these results, the values obtained for k are $0.37 \pm .04$ from M_f and $0.30 \pm .04$ from M_m. Within the observational uncertainties k is consistent with the theoretical value of 0.30 derived from a range of seismic models (Farrell 1972). It should be noted that there are also oceanic contributions to the M_m and M_f terms which have not been taken into account. A least-squares analysis of the data for the 13.633 day term gave effectively zero amplitude as compared with the theoretical value of 0.3 ms.

CONCLUSIONS

20 years of PZT time and latitude observations have been analysed for the principal lunar semi-diurnal constituent M_2. Within the experimental and theoretical uncertainties these observations are consistent with a theoretical model including both the body tide and ocean tides.

Analysis of the time observations for the long-period tides M_m and M_f gives amplitudes which, within the experimental uncertainties, are consistent with the expected value for the body-tide Love number k.

REFERENCES

1. Baker, T.F. 1977 Earth Tides, Crustal Structure and Ocean Tides. 8th International Symposium on Earth Tides, Bonn.
2. Farrell, W.E. 1972 Deformation of the Earth by Surface Loads. Reviews of Geophysics and Space Physics $\underline{10}$, 761-797.
3. Melchior, P. 1978 The Tides of the Planet Earth. Pergamon Press.
4. O'Hora, N.P.J. 1973 Semi-Diurnal Tidal Effects in PZT Observations. Physics of the Earth and Planetary Interiors $\underline{7}$, 92-96.
5. O'Hora, N.P.J. and Griffin, S.F. 1977 Short-period Terms in Time and Latitude Observations made with the Herstmonceux Photographic Zenith Tube. Proceedings of IAU Symposium No.78 (in press).
6. Thomas, D.V. and Wallis, R.E. 1971 Results obtained with a Danjon Astrolabe at Herstmonceux. R.Obs.Bull.No.160.
7. Vondrák, J. 1969 A Contribution to the Problem of Smoothing Observational Data. Bull. of the Astron. Inst. of Czechoslovakia 20, 349.

POLAR MOTION AND EARTH ROTATION MONITORING IN CANADA

J. Popelar
Earth Physics Branch, Department of Energy, Mines and
Resources, Ottawa, Canada.

PZT observations of the Earth's rotation and polar motion have been carried out from observatories near Ottawa and Calgary since 1952 and 1968 respectively. A comprehensive re-evaluation and analysis of the PZT data is currently under way, using a new self-contained computer program which can easily accommodate changes of astronomical constants as well as modifications of the star catalogue.

So far Ottawa PZT observations from the period 1956-1977 have been re-reduced in a uniform manner. Insufficient documentation and poor data quality made the re-processing of the 1952-1955 observations unprofitable. These results were obtained using two instruments (PZT 1 during 1956-1968 and PZT 2 since March 1968) operating from four different locations (the PZT was relocated in March 1960, January 1970 and December 1970). The interruptions in the observation series have been removed using a least squares technique for determining the datum bias and linear trend.

The observed rotational time (UT0) has been referred to an atomic time system which corresponds to TAI since 1961; the observations prior to 1961 have been related to TAI using A.1 (USNO). No appreciable linear trend has been determined for the latitude observations over the 22 years although some long period (> 1200 days) variations with relatively small amplitudes (< $0\rlap{.}''03$) can be detected. The observations of rotational time show an average offset of -247×10^{-10} with respect to the atomic time scale over the 22 year period. With some fluctuation the offset has a clear tendency to decrease by about -12×10^{-10} per year, from 101×10^{-10} in 1956 to -321×10^{-10} in 1977; significant deviations from this average annual change occurred between 1963-64 (-58×10^{-10}) and 1973-74 ($+31 \times 10^{-10}$).

A preliminary least squares harmonic analysis positively detected the Chandler motion in both the latitude and UT0 observations as being very closely represented by a single harmonic with the amplitude of $0\rlap{.}''135$ and the period of 434 days. The analysis of seasonal variations proved to be more complex and has not yet been completed. Due to a number of factors (e.g. systematic errors in the star catalogue, irregularities in the Earth's rotation) the seasonal variations cannot be satisfactorily represented by a simple superposition of two harmonics (annual and semi-annual).

The Calgary PZT observations will also be re-reduced in a similar way; since the Calgary PZT was set up at the latitude of the Herstmonceux PZT of the Royal Greenwich Observatory in England and the same star catalogue is used, detailed analysis and comparison of the results obtained with the two instruments will follow.

Since 1974 two satellite Doppler tracking stations have been installed at the Ottawa and Calgary PZT observatories to compare polar motion as derived from the astronomical and satellite techniques and to analyse the station-related effects in the two independent systems. The satellite results have not shown apparent seasonal variations of the stations' coordinates; however, more data are needed to carry out a satisfactory harmonic analysis and assess the long term stability of the satellite system.

The Earth Physics Branch is currently sponsoring a planned experiment to use a long baseline radio interferometer for regular observations of the Earth's rate of rotation. Using the nearly east-west 3075 km baseline between the Algonquin Radio Observatory at Lake Traverse, Ontario, and the Dominion Radio Astrophysical Observatory in Penticton, British Columbia, UT1 is to be determined at regular intervals during the summer of 1979. The plans call for operation of the interferometer in a phase-stable mode by means of pilot tones transmitted via the Canadian communications satellite ANIK-B. The preliminary error budget indicates that atmospheric effects will limit the accuracy of UT determinations in this experiment to about 1-2 ms.

The main objective of this effort is to provide an accurate reference frame for world-wide precise geodetic positioning and for studies of the dynamics of the Earth. The general recognition of the dynamic evolution of the Earth emphasizes the need for a diversified, technologically advanced monitoring service which has, besides its scientific significance, important practical applications for resource development and for studies of geophysical effects on the Earth's environment. Continuity and detailed comparison of the results from various techniques during periods of simultaneous operation are essential to full understanding of the physical phenomena involved and to the proper reduction and interpretation of the results of the observations.

A high degree of automation of routine operating procedures, data processing and communication by means of a mini-computer network greatly increases the efficiency of this long-term program; it also expedites the activities of data management and exchange which are essential, because of the need for wide-ranging international cooperation, in studies of polar motion and the rotation of the Earth.

POLAR COORDINATES AND UT1 - UTC FROM PZT OBSERVATIONS

D. Djurovic
Department of Astronomy
Belgrade, Yugoslavia

1. INTRODUCTION

Use of new techniques (Doppler, laser, VLBI) has yielded such results that many astronomers believe that in the near future these techniques will replace classical instruments for observations of the Earth's rotation. None of the modern techniques has furnished observational series which demonstrate that systematic errors in the polar coordinates and UT1 are stable over sufficiently long intervals. Investigations of known phenomena such as sudden changes of the secular term in UT1 - UTC (Munk and MacDonald, 1960), changes of the amplitude and phase of the seasonal irregularities (Fliegel and Hawkins, 1967; Pavlov and Staritzin, 1962), secular motion of the mean pole (Mihailov, 1971; Markowitz, 1960), continental drift (Stoyko, 1938; Djurovic, 1976), quasi-diurnal nutation of Molodenskij (Popov, 1963; Toomre, 1974; Rochester et al., 1974), etc., are complicated by the existence of systematic error variations (accuracy), and to a smaller degree, on the accidental errors (precision). The BIH and IPMS make use of individual series of astronomical latitude, ϕ_i, and (UT0 - UTC)$_i$ from up to 82 classical instruments (Guinot, 1976). If the systematic errors in these series are independent, the polar coordinates and UT1 - UTC determined from a combination of observational series would result in an improvement of accuracy by at least one order of magnitude through the mutual compensation of the variation in systematic errors. At present, however, it is uncertain whether mutual independence of modern observations (a basic condition for mutual compensation of errors) will be better than that of the classical instruments.

Accordingly, the continuation of parallel observations by new and classical techniques remains indispensable as long as there is no definite proof that x, y, and UT1 - UTC obtained by the new techniques are more homogeneous than the results achieved by classical methods.

2. POLAR COORDINATES AS OBTAINED BY THREE TYPES OF INSTRUMENTS

Observational data used for the computation of x and y are the instantaneous latitudes, ϕ_i, observed in the period 1967-1971 and communicated to the BIH by 42 observatories. Table A of a previous paper (Djurovic, 1975a) lists the observatories considered. The polar coordinates are computed according to methods described previously (Djurovic, 1975a; Djurovic, 1975b). The only difference is that the computations are carried out separately for three types of instruments (astrolabe - A, photographic zenith tube - PZT, and zenith telescope - lz).

Yearly means of the differences of our results and the corresponding results of the IPMS, smoothed using Vondrak's method, $\Delta x = x - x(IPMS)$ and $\Delta y = y - y(IPMS)$ are given in Table 1. Variations of the differences, $\Delta x(PZT)$ and $\Delta y(PZT)$, appear somewhat larger than those of the other two types of instruments. This is explained by the fact that the number of PZTs (8) are lower than that of the astrolabes and zenith telescopes. With 15 PZTs in operation in 1975 it is surely possible to attain a stability of $\pm 0\overset{''}{.}01$. The same conclusion results from the analysis of the stability of the reference system by the Wilcoxon method (Djurovic, 1978).

The second important condition satisfied by x(PZT) and y(PZT) is that their deviations from x(IPMS) and y(IPMS) do not contain periodic terms with periods close to that of the Chandler or annual period as, for example, $\Delta x = x(BIH) - x(IPMS)$. Standard deviations ε_x and ε_y for the PZT solutions are somewhat less than those for the BIH coordinates, but it should be kept in mind that the former are associated with 20-day intervals while the latter are related to 5-day intervals. The values for the period 1967-1971 are:

$\varepsilon_x(PZT) = \pm 0\overset{''}{.}015$, $\qquad \varepsilon_x(BIH) = \pm 0\overset{''}{.}017$,
$\varepsilon_y(PZT) = \pm 0.014$, $\qquad \varepsilon_y(BIH) = \pm 0.018$.

3. THE ACCURACY OF THE MEAN SYSTEM $(UT1 - UTC)_{PZT}$

Observational material used in the present analysis is from the observations of UT0 - UTC at 56 observatories reported to the BIH for the period 1967-1974. All details of the data processing and of the mean system $(UT1 - UTC)_u$ are set forth in a previous paper (Djurovic, 1975b). For the observatories H, G, MS, MCP, N, TO, and W, whose series are uninterrupted in the interval 1967-1974 we have computed a system of differences,

$$d_{ij} = (UT1 - UTC)_{ij} - (UT1 - UTC)_u,$$

where $(UT1 - UTC)_{ij}$ is the result of observatory, i, on the night of observation, j. From 1971-1974 three more observatories, CL, OS, and RCP, were included. Mean values of d_{ij}, denoted by D_i, have been calculated for 30-day intervals for each observatory, i. Two mean systems of D_i were calculated: for the group of the first 7 observatories (Z), and for the group of all 10 observatories (Z'). The results obtained reveal that Z and Z' are varying within the limits of a few milliseconds.

They do not contain periodic components with periods close to the annual or semiannual terms which would impede the study of the seasonal irregularities in the Earth's rotation.

Table 1. Differences in the polar coordinates for different types of instruments.

	$\Delta x(A)$	$\Delta x(PZT)$	$\Delta x(LZ)$	$\Delta y(A)$	$\Delta y(PZT)$	$\Delta y(LZ)$
1967	-0".049	-0".032	-0".030	0".055	0".025	-0".009
1968	-0.044	-0.075	-0.016	0.056	0.034	-0.023
1969	-0.046	-0.086	-0.022	0.077	0.037	-0.014
1970	-0.047	-0.058	-0.010	0.088	0.062	-0.033
1971	-0.033	-0.034	-0.014	0.071	0.089	-0.026
mean	-0.044	-0.057	-0.018	0.069	0.049	-0.021

Accidental errors were estimated by the expressions:

$$\sigma^2 = \frac{1}{2(N-1)} \sum_{k=2}^{N} (Z_k - Z_{k-1})^2, \text{ and}$$

$$\varepsilon_i^2 = \frac{\sum_j p_{ij}(d_{ij} - D_i)}{\sum_j p_{ij}},$$

where k refers to the 30-day interval, and p_{ij} is the weight of d_{ij}. The results obtained for σ and ε (mean of ε_i) are $\sigma = \pm 0^s.0013$ (the corresponding values for Z' is $\sigma' = \pm 0^s.0011$), and $\varepsilon = \pm 0^s.0072$. Standard deviations of $(UT1 - UTC)_u$ (resolution interval is one day) and $(UT1 - UTC)_{BIH}$ (resolution interval of five days) are $\sigma_u = \pm 0^s.0027$ (Djurovic, 1975b), and $\sigma_{BIH} = \pm 0^s.0018$ (Guinot, 1976). From the comparison of σ and ε with σ_u and σ_{BIH} it appears that $(UT1 - UTC)_{PZT}$ is equivalent to the mean systems of $(UT1 - UTC)_u$ and $(UT1 - UTC)_{BIH}$ with respect to accidental errors.

4. CONCLUSION

With the present number of PZTs the accuracy of the BIH five-day mean of polar coordinates is greater than that obtained from a solution using PZTs only. The present accuracy can be attained with a PZT solution if a longer averaging period were used (approximately ten days).

As individual PZT systems of $(UT0 - UTC)_{ij}$ are more homogeneous than $(UT0 - UTC)_{ij}$ obtained from the observations of transit instruments and astrolabes (Djurovic, 1978), the mean system $(UT1 - UTC)_{PZT}$ is equivalent to the $(UT1 - UTC)_{BIH}$. The continuity of the classical observations of the Earth's rotation can be preserved by existing PZTs only.

REFERENCES

Djurovic, D.: 1975a, Bull. Obs. Astron. Belgrade 126, p. 62.
Djurovic, D.: 1975b, Bull. d'Observations, Obs. Royal Belgium 4, fasc. 3, p. 17.
Djurovic, D.: 1976, Bull. Accad. Serbe Sc. et Arts 55, no. 9, p. 7.
Djurovic, D.: 1978, Publ. Depart. Astron. Belgrade 8, p. 5.
Fliegel, H. F., and Hawkins, T. P.: 1967, Astron. J. 72, p. 544.
Guinot, B.: 1976, Report presented to IAU General Assembly, Grenoble.
Markowitz, W.: 1960, in S. K. Runcorn (ed.), "Methods and Techniques in Geophysics", Intersciences Publishers Ltd., London.
Mihailov, A. A.: 1971, Astron. Zhurn. Akad. Naouk SSSR 48, no. 6, p. 1301.
Munk, W. H., and MacDonald, G. J. F.: 1960, "The Rotation of the Earth", Cambridge University Press, London.
Pavlov, N. N., and Staritzin, G. V.: 1962, Astron. Zhurn. Akad. Naouk SSSR 39, no. 1, p. 123.
Popov, N. A.: 1963, Astron. Zhurn. Akad. Naouk SSSR 7, p. 422.
Rochester, M. G., Jensen, O. G., and Smylie, D. E.: 1974, Geophys. J. Roy. Astron. Soc. 38, p. 349.
Stoyko, N.: 1938, Acta Astron. 3, p. 97.
Toomre, A.: 1974, Geophys. J. Roy. Astron. Soc. 38, p. 335.

(α-2L) TERMS AS OBTAINED FROM PZT OBSERVATIONS

Shigetaka Iijima, Shigeru Fujii, and Yukio Niimi
Tokyo Astronomical Observatory, University of Tokyo

Abstract

Periodic terms with arguments such as (α-2L) and (2α-2L) etc. which appear via various phenomena in the results of time and latitude observations by PZT are summarized in analytical forms in order to clarify their mutual relation. Numerical values in time and latitude for these terms to be expected at the Tokyo PZT are also given. From the data of time observations obtained by the Tokyo PZT during the past 15 years, amplitudes of the terms, $\cos(\alpha-2L_\mathrm{C})$ and $\sin(2\alpha-2L_\mathrm{C})$ in time, are obtained as +0.72 and 1.21 ms, respectively. These results give the value of $(1+k-\ell)$ as 1.14, and the values of nutation coefficients for $2L_\mathrm{C}$, $\Delta\varepsilon=+0''.0893$ and $\sin\varepsilon\cdot\Delta\Psi=-0''.0818$.

1. Relations among α, 2α, (α-2L), and (2α-2L) terms

Periodic terms in time and latitude with the arguments of α, 2α, (α-2L), and (2α-2L) caused by the effects of the Moon or the Sun appear in the results of PZT observations via four kinds of phenomena. In these terms, α refers to the right ascension of the stars observed, i.e. the local sidereal time of the observation, and L the mean longitude of the Moon or the Sun. These phenomena are as follows:

1) deflection of the plumb line,
2) deformation by the Earth tide,
3) forced polar motion by the external torque, (Oppolzer terms)
4) errors in adopted nutation coefficients.

Among these, 1) appears as a systematic variation in the adopted longitude or latitude, 2) and 3) appear as the longitude and latitude variations as a result of the forced polar motion, and 4) appears in the star positions to which the PZT observations are referred. These periodic terms in observed time and latitude, δt and $\delta\phi$, due to each phenomenon are tabulated in analytical forms in Table 1.

Table 1. Relation among α, 2α, $(\alpha-2L)$, and $(2\alpha-2L)$ terms.

δt / $\delta\phi$		cos/sin α	sin/cos 2α	cos/sin $(\alpha-2L_{\mathbb{C}})$	cos/sin $(\alpha-2L_{\odot})$	sin/cos $(2\alpha-2L_{\mathbb{C}})$	sin/cos $(2\alpha-2L_{\odot})$
$\dfrac{3}{2}\dfrac{G}{\omega^2}\left(\dfrac{M}{r^3}\right)\mathit{d}$ / $\dfrac{1000}{15}\,\text{ms}$, $1''$	$\dfrac{C-A}{C}$ / $\dfrac{C}{C-A}\times m$	$(\cos i - \sin^2 i + \zeta)$ $\times \sin\varepsilon \cos\varepsilon$	$\times \dfrac{\sin^2\varepsilon}{2}$	$\cos i$ $\times \sin\varepsilon\,\dfrac{1+\cos\varepsilon}{2}$	ζ $\times \sin\varepsilon\,\dfrac{1+\cos\varepsilon}{2}$	$\cos i$ $\times \left(\dfrac{1+\cos\varepsilon}{2}\right)^2$	ζ $\times \left(\dfrac{1+\cos\varepsilon}{2}\right)^2$
Deflection of plumb line	$(1+k-\ell)$ / $\dfrac{C}{C-A}\times m$ — $\tan\phi\cos 2\phi$ / $1\;\sin\phi\cos\phi$	-1	0	1	1	0	0
Deformation by earth tide	$k\left(\dfrac{m'}{3J_2}\right)$ — 1	0	1	0	0	1	1
Forced p.m. by ext. torque	1 — $\tan\phi$ / 1	∓ 1	0	$\mp 2\,\dfrac{n_{\mathbb{C}}}{\omega}\bigg/\left(1-2\dfrac{A}{C}\right)$	$\mp 2\,\dfrac{n_{\odot}}{\omega}\bigg/\left(1-2\dfrac{A}{C}\right)$	$\pm 1\,\dfrac{A}{C}\dfrac{n_{\mathbb{C}}}{\omega}\bigg/\left(1-2\dfrac{A}{C}\right)$	$\pm 1\,\dfrac{A}{C}\dfrac{n_{\odot}}{\omega}\bigg/\left(1-2\dfrac{A}{C}\right)$
Errors in nuta. coef.	ξ — $\tan\phi$ / 1	0	0	$\mp\dfrac{\omega}{2n_{\mathbb{C}}}$	$\mp\dfrac{\omega}{2n_{\odot}}$	0	0

The terms in δt and $\delta\phi$ are shown in a composite form in Table 1 to display the mutual relationship at a glance. Quantities devided by slanting lines and/or accompanied by double signs are to be read separately following the arranged order. That is, the left side of slanting lines and the upper sign in double signs refer to δt, and the right side and the lower sign to $\delta\phi$. For instance, the $(\alpha-2L)$ term in δt attributed to the deflection of plumb line must be read as follows:

$$\frac{3}{2}\frac{G}{\omega^2}\left(\frac{M}{r^3}\right)_{\text{☾}}(1+k-\ell)m\cdot\cos i\cdot\sin\varepsilon\frac{1+\cos\varepsilon}{2}\tan\phi\frac{1000^{\text{ms}}}{15}\sin(\alpha-2L_{\text{☾}}).$$

The following notation is used in Table 1:

- i : inclination of the Moon's orbit to ecliptic,
- ε : obliquity of the ecliptic,
- M : mass of the heavenly body shown by the suffix,
- r : geocentric mean distance of the Moon or the Sun,
- G : constant of gravitation,
- ω : Earth's angular speed of rotation,
- n : mean motion of the Moon or the Sun,
- k : Love number,
- ℓ : Shida number,
- ϕ : latitude of the observing site,
- $m = \omega^2 a/g_e$,
- a : equatorial radius of the Earth,
- g_e : acceleration of gravity on the equator,
- $m' = \omega^2 a^3/(GM_\oplus)$,
- $J_2 = (C-A)/(M_\oplus a^2)$,
- $\zeta = (M/r^3)_\odot/(M/r^3)_{\text{☾}}$,
- $1+\xi$: ratio of adopted to true values of nutation coefficients,
- A, A, C : principal moments of inertia of the Earth.

It should be noticed in this table that:

a) δt and $\delta\phi$ in each column of periodic terms are expressed solely by a sine or cosine, respectively, regardless of the phenomena,
b) 2α and $(2\alpha-2L)$ terms appear only via phenomenon 1),
c) the amplitude ratio of $(\alpha-2L)$ to $(2\alpha-2L)$ terms attributed to 1) is $2\tan\phi\sin\varepsilon/(1+\cos\varepsilon)$ for δt, and $4\cot 2\phi\sin\varepsilon/(1+\cos\varepsilon)$ for $\delta\phi$, regardless of Love number,
d) α and 2α terms must appear as a part of systematic errors in the star positions,
e) the amplitude of the periodic terms via 2) is negligiblly small as compared with those via 3) because of the existence of the factor of n/ω in 2),

f) $(\alpha-2L)$ terms as obtained from PZT observations appear as an ensemble of the terms via 1) to 4). However, the one via 1) can be separated by use of the result of $(2\alpha-2L)$ terms as well as the amplitude relation mentioned in c).

g) $mC/(C-A)$ and $m'/(3J_2)$ in Table 1 are almost equal to unity.

Amplitudes of α, 2α, $(\alpha-2L)$, and $(2\alpha-2L)$ terms in Table 1 are evaluated for the site of the Tokyo PZT, and tabulated in Table 2, to two significant figures after the decimal point, the last figure of which is rounded.

Table 2. $\delta t/\delta\phi$ expected at the Tokyo PZT.

ms 0".01	cos/sin α	sin/cos 2α	cos/sin $(\alpha-2L_\text{\it d})$	cos/sin $(\alpha-2L_\odot)$	sin/cos $(2\alpha-2L_\text{\it d})$	sin/cos $(2\alpha-2L_\odot)$
Defl. of plumb line	−0.53 / −0.36	+0.16 / +0.11	+0.38 / +0.25	+0.18 / +0.12	+1.28 / +0.91	+0.59 / +0.42
Deform. by earth tide	0	0	−0.00 / +0.00	−0.00 / +0.00	0	0
Forced p.m. by torque	−0.42 / +0.87	0	+0.32 / −0.67	+0.14 / −0.29	0	0
Errors in nut.coef.	0	0	+0.04 / −0.08	+0.25 / −0.53	0	0
Σ	−0.94 / +0.51	+0.16 / +0.11	+0.74 / −0.49	+0.56 / −0.70	+1.28 / +0.91	+0.59 / +0.42

Numerical values of the relevant constants adopted in this table are as follows:

$m = 0.003\,4678$,
$m' = 0.003\,4614$,
$J_2 = 0.001\,0826$,
$\zeta = 0.45924$,

$\phi = 35°40'20".707$,
$1+k-\ell \equiv 1.20$,
$k \equiv 0.28$,
$\xi \equiv -0.01$.

Quantities in Table 2 divided by slanting lines are to be read separately for δt and $\delta\phi$, respectively, following the arranged order, as in the preceding table.

2. $(\alpha-2L_d)$ and $(2\alpha-2L_d)$ terms in δt as obtained from the PZT observations in Tokyo.

Observational data obtained by the Tokyo PZT during 15 years from 1962 through 1976 are analyzed here. The time data amounting to about 48,000 star observations made during the 15 years are reduced to the consistent system on the PZT star catalog, α_{75}, which has been used since 1975, allowing for all the past changes in the adopted longitude and the aberration constant etc. back to 1962. (UT0-UTC) for each star observation is converted to (UT1-UTC) by use of the results of the polar coordinates of the ILS and the IPMS, and then transferred to (UT1-TAI) by using the (UTC-TAI) data of the BIH.

In order to remove the seasonal and long term variations contained in the observed UT1, an observational equation composed of annual, semi-annual, biennial terms, and a cubic equation with respect to time is applied to the data of (UT1-TAI) by the method of least squares. This fitting process is repeated for every four-year interval of data, shifting by three-year steps. The residual, (O-A), over three years in the middle of each four-year interval are used for the subsequent calculation.

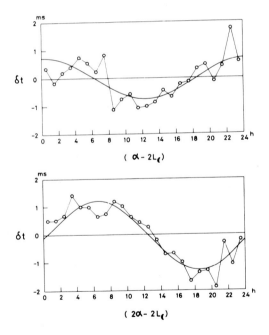

Fig. 1. Results of $(\alpha-2L_d)$ and $(2\alpha-2L_d)$ terms.

The data of (O-A) for each star observation are classified with respect to the phase of $(\alpha-2L_{\mathcal{C}})$ or $(2\alpha-2L_{\mathcal{C}})$ into one-hour intervals such as 0h - 1h, 1h - 2h, ..., 23h - 0h. The hourly mean values of (O-A) in phase are illustrated in Figure 1 for $(\alpha-2L_{\mathcal{C}})$ and $(2\alpha-2L_{\mathcal{C}})$ terms in δt. Each small circle in these graphs is the mean value of about 2,000 star observations. The results shown in Figure 1 are given analytically as follows:

$$\delta t = \underset{\pm.99}{0.725}^{\text{ms}} \cos(\alpha-2L_{\mathcal{C}}) + \underset{\pm.98}{0.071}^{\text{ms}} \sin(\alpha-2L_{\mathcal{C}}), \qquad (1)$$

$$\delta t = \underset{\pm.75}{1.213} \sin(2\alpha-2L_{\mathcal{C}}) - \underset{\pm.76}{0.169} \cos(2\alpha-2L_{\mathcal{C}}). \qquad (2)$$

The amplitude of the cosine term of $(\alpha-2L_{\mathcal{C}})$ in (1) and that of the sine term of $(2\alpha-2L_{\mathcal{C}})$ in (2) are in rather good agreement with the corresponding values in Table 2. By comparing the result of 1.213 ms obtained for the $\sin(2\alpha-2L_{\mathcal{C}})$ term in (2) with the corresponding value in Table 2, the value of $(1+k-\ell)$ can be estimated as 1.14. The amplitude of 1.213 ms for the $\sin(2\alpha-2L_{\mathcal{C}})$ term in (2) also gives the amplitude of the $\cos(\alpha-2L_{\mathcal{C}})$ term due to the deflection of plumb line as 0.361 ms, by applying the ratio of $2 \tan \phi \sin \varepsilon/(1+\cos \varepsilon) = 0.2978$. After removing 0.361 ms just obtained and 0.322 ms due to 3) in Table 2 from the ensemble result of 0.725 ms for $\cos(\alpha-2L_{\mathcal{C}})$ term in (1), 0.042 ms is obtained as the residual which corresponds to the term ascribable to the errors in nutation coefficients for $2L_{\mathcal{C}}$. Thus the results, $\Delta\varepsilon$ and $\sin \varepsilon \cdot \Delta\Psi$, are obtained as follows:

	$\Delta\varepsilon$	$\sin \varepsilon \cdot \Delta\Psi$
Tokyo	$0\overset{''}{.}0893 \pm 0\overset{''}{.}0022$,	$-0\overset{''}{.}0818 \pm 0\overset{''}{.}0022$.

McCarthy (1976) obtained similar results from the data of PZT observations made at Washington, Richmond, and Herstmonceux, assuming the value of $(1+k-\ell)$ equal to 1.13. These results are shown here for comparison:

	$\Delta\varepsilon$		$\sin \varepsilon \cdot \Delta\Psi$	
Washington	$0\overset{''}{.}0907$	$\pm\ 0\overset{''}{.}0016$,	$-0\overset{''}{.}0832$	$\pm\ 0\overset{''}{.}0015$,
Richmond	977	08,	897	07,
Herstmonceux	889	17,	816	16.

The present results are rather close to those obtained from Herstmonceux data by McCarthy.

Reference

McCarthy, D.D.: 1976, Astron. J. 81, p. 482.

ETUDE SUCCINTE DU CATALOGUE FONDAMENTAL FK4 A PARTIR
DES OBSERVATIONS FAITES A L'ASTROLABE

G. Billaud
Centre d'Etudes et de Recherches Géodynamiques et
Astronomiques - Grasse (France)

ABSTRACT

A short study of FK4 catalogue is based on CGA (Billaud G. et al, 1978). Important differences are shown in right ascension as well as in declination and assumptions are made to explain the discrepancies between both catalogues.

1. INTRODUCTION

Le Catalogue Général des Astrolabes (CGA) qui sert de base à cette étude (Billaud G. et al, 1978) a été réalisé à partir de 20 catalogues élémentaires indépendants observés à l'astrolabe de 1958 à 1975. C'est un catalogue relatif qui fournit les différences (Astrolabe-FK4) pour 1139 étoiles fondamentales régulièrement réparties entre les déclinaisons -63 et +79° avec une précision moyenne de 0,004s en α et de 0",06 en δ.

2. ETUDE EN ASCENSION DROITE

L'étude porte sur 597 étoiles de l'hémisphère nord et 531 étoiles de l'hémisphère sud. Le découpage en régions de une heure en ascension droite montre un écart $\Delta\alpha_\alpha$ pseudo-périodique d'amplitude totale d'environ 6ms, dont l'aspect est indépendant de l'hémisphère considéré, avec deux maxima à 6 et 18 heures. L'analyse des quantités $\Delta\alpha_\delta$ par bande de 15° de déclinaison confirme l'important effet déjà observé dans l'hémisphère sud (Anguita C. et Noël F., 1969).

L'aspect topographique de la répartition est caractérisé par une vallée est-ouest dans l'hémisphère nord et une pente abrupte dans l'hémisphère sud. Ces deux régions communiquent entre elles par deux cols situés à 0 et 12 heures, cols qui isolent deux "collines" correspondant aux maxima.

3. ETUDE EN DECLINAISON

L'analyse en $\Delta\delta_\alpha$ porte sur 524 étoiles dans l'hémisphère nord et 422 étoiles dans l'hémisphère sud. Malgré une dispersion plus importante dans l'hémisphère sud, due aux grands écarts individuels, l'aspect des $\Delta\alpha_\delta$ est le même des deux côtés de l'équateur et montre deux maxima séparés de 12 heures et correspondant à une variation d'amplitude totale d'environ 0",1.

Quant aux $\Delta\delta_\delta$ il faut noter une variation négative de 0",06 au delà de $\delta = 45°$ quand on va vers le pôle et une anomalie positive importante (0",15) liée à une forte dispersion dans la zone -45°, -30°. Cette dispersion s'explique par la présence, entre -30 et -35° d'étoiles dont les écarts élevés atteignent $\pm 1"$. De -35 à -45° $\Delta\delta_\delta = 0",25$.

L'aspect topographique de la répartition des $\Delta\delta$ est beaucoup plus tourmentée que celle des $\Delta\alpha$. Le trait le plus remarquable est le voisinage, dans l'hémisphère sud de reliefs et de creux correspondant à des écarts supérieurs à 0",4 en moyenne.

4. CONCLUSIONS

Les différences signalées entre CGA et FK4 sont importantes et significatives. A priori il est impossible de les attribuer à l'un ou l'autre des catalogues. On remarquera toutefois que les observations à l'astrolabe procèdent d'une manière toute différente de celles des cercles méridiens sur lesquels repose le FK4. En particulier, à l'astrolabe, on peut observer quasi simultanément des étoiles différant de près de 60° en déclinaison et de 4 heures en distance zénithale, ce qui aurait pour effet de répartir sur la sphère une éventuelle erreur systématique liée à la date. La méthode utilisée est telle, d'autre part, que deux régions situées au nord et au sud du zénith sont observées avec la même précision. Enfin l'accord des $\Delta\alpha_\delta$ dans l'hémisphère sud avec des observations méridiennes récentes montrent que les catalogues d'astrolabes ne présentent pas d'écarts systématiques importants et que l'on a lieu de penser que les différences signalées sont réelles. S'il en est ainsi, l'astrolabe peut effectivement apporter une forte contribution à la révision du catalogue fondamental FK4.

REFERENCES

Billaud G., Guallino G., Vigouroux G., 1978, Astron. Astrophys. 63, pp. 87-95

Anguita C. and Noël F., 1969, Astr. J. 74, pp. 954-957

DISCUSSION

P. Pâquet: Pouvez-vous préciser l'origine de la pseudo-périodicité observée dans les différences entre les catalogues "astrolabe" et "méridien"?

G. Billaud: Il s'agit de la différence entre astrolabes et FK4. A priori on ne peut pas dire si l'effet périodique appartient aux astrolabes ou aux méridiens sur lesquels repose le catalogue FK4. Cependant, le fait que les observations $\Delta\alpha_\delta = \Delta\alpha - \Delta\alpha_\alpha$ sont confirmées par des observations méridiennes (Anguita, C. et Noël, F., 1969) laisse à penser que les astrolabes sont dans le vrai. On peut aussi remarquer que les observations sont menées de manière différente avec les deux types d'instruments: "chronologique" au méridien, tandis que l'astrolabe observe quasi simultanément deux étoiles dans une zone de $60°$ en δ et de 5 heures en α.

PART II : POLAR MOTION

ON THE COORDINATE SYSTEMS USED IN THE STUDY OF POLAR MOTION

E. P. Fedorov
Main Astronomical Observatory
Ukrainian Academy of Sciences
Kiev, U.S.S.R.

IAU Symposium No. 78 "Nutation and the Rotation of the Earth" held in Kiev in 1977 revealed a certain lack of precision in the fundamental concepts and some looseness of terminology employed in the treatment of this problem. When talking about polar motion we should give, first of all, rigorous conceptual definitions of both the pole and a reference frame in which it moves. The selection of a reference system was the topic of an IAU Colloquium held in Torun in 1974. Although the discussion there was thorough and comprehensive, it did not result in the removal of all ambiguities which have tarnished discussion of the problems in the understanding of the Earth's rotation.

"Since distances are not directly measured in classical astronomy but have to be inferred by indirect methods, the systems of coordinates in common use are those that specify only directions" (Clemence, 1963). Since direction is completely given by a pair of angular coordinates, the space in which such systems are realized is a two-dimensional space. To emphasize this point Brandt (1975) suggested that astronomers may make use of the terminology employed in geometrical optics, where the real three dimensional space in which physical bodies are located is called the object space and the space in which images of these bodies are located is called the image space. Three dimensional coordinate systems in the object space may be either inertial or non-inertial, but the very conception of "inertiality" has no meaning for systems realized in the image space. In that case we may only discuss rotating and non-rotating systems. The image space used in astronomy is the two dimensional surface of a unit sphere. Recently, however, some authors have used other representations (Fedorov, 1976a; Zhongolovitch, 1977; Murray, 1978).

Assume that we have in the object space several direct lines connecting some points of different celestial bodies (Figure 1). Let us take in the image space an arbitrary point, 0, and draw from it unit vectors, $\bar{s}_i$, parallel to these lines. Obviously, several parallel lines in the object space will be represented by only one vector in the image space.

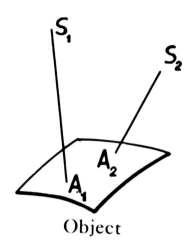

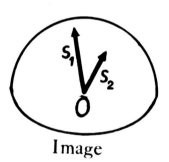

Figure 1. Object and image spaces.

Directions from the observer to extragalactic sources are practically independent of the observer's motion on the Earth, together with the Earth, or in space about the Earth (assuming that correction for aberration is taken into account). The proper motions of extragalactic sources also being negligible, the unit vectors along the directions from the observer to the object can be assumed to be fixed in the image space. On the other hand, we have the following directions linked to the Earth and rotating with it: geocentric position vectors of the points on the Earth's surface, plumb lines at these points, chords connecting these points (baselines of radio interferometers). We may draw from O in the image space unit vectors, $\bar{e}_j$, parallel to these directions.

Astronomical observations of stars, planets, the Moon, artificial satellites, or radio sources enable the rotation of the pencil of vectors, $\bar{e}_j$, with respect to the vectors, $\bar{s}_i$, to be monitored. Measurement of either the angles or distances (with laser ranging techniques) are capable of providing the necessary data (Fedorov, 1976b). These data would comprise sufficient information on the rotation of the Earth so that this phenomenon could be studied without the use of coordinates (Veis, 1976), but the use of coordinates simplifies analysis. Since the angles between the unit vectors, $\bar{s}_i$, are practically constant, a celestial coordinate system, XYZ, can be rigidly linked to them. The following resolution concerning this matter was adopted at the Toruń Colloquium:

> The celestial system will be defined by a catalog of adopted conventional coordinates of extragalactic sources. These coordinates could be obtained from the best available observations and reduced to a given epoch in the existing celestial system (FK4 or FK5). After such a catalog is constructed and adopted, reference to the original celestial system may be dropped. Further improvements in the realization of the system would come through the compilation of better catalogs of extragalactic sources (e.g. with no reference to any plane or direction pertaining to the Earth or Solar System).

It is quite natural that a three dimensional reference coordinate system in the object space is needed for description of various phenomena dealt with in Earth dynamics. However, to study the Earth's rotation one can use systems realized in the same two dimensional space as the non-rotating system, XYZ. Then relative orientation of the terrestrial and celestial systems would be defined only by the angles between their axes. The phenomenon called the rotation of the Earth is, in essence, the variation of these angles.

Any terrestrial system in the image space may be attached to the pencil of the unit vectors $\bar{e}_j$, but not rigidly since these vectors (in the case of a non-rigid Earth) do not maintain their directions relative to one another. For the non-rigid attachment to be realized, certain conditions should be imposed on the relationship of a system, xyz, with respect to the vectors $\bar{e}_j$. Such a system may be called the conventional terrestrial system. Its rotation is not exactly predictable since it is affected by excitation by some geophysical processes. So it is convenient to introduce an intermediate system, $\xi \eta \zeta$, whose rotation approximates as close as possible that of the system, xyz, and at the same time is precisely predictable. It may be called the terrestrial ephemeris system. This system as well as the system, xyz, can be transformed into the non-rotating celestial frame by means of the equations:

$$(X,Y,Z) = M_0 (\xi, \eta, \zeta) = M (x,y,z), \tag{1}$$

where M_0 is the matrix of precession, nutation, and the "ephemeris" diurnal rotation of the Earth. We may write

$$M = M_0 (I + \sigma), \tag{2}$$

where I is the unit matrix, and

$$\sigma = \begin{vmatrix} 0 & -w & v \\ w & 0 & -u \\ -v & u & 0 \end{vmatrix}, \qquad (3)$$

u, v, w being small rotation angles about the x, y, z-axes respectively (Fedorov et al., 1972). From (2) we have

$$M - \underline{M} = M_0 \sigma. \qquad (4)$$

It should be noted that the elements of the matrix change with nearly diurnal periods.

Proceed now to possible realizations of different coordinate systems, and consider first the selection of the ζ-axis of the ephemeris terrestrial system. For this axis the following directions may be adopted.

1. The axis of the total angular momentum of the Earth, $\overline{H}$

We donote the unit vector parallel to $\overline{H}$ by $\overline{h}$. The motion of $\overline{h}$ with respect to the non-rotating system XYZ is independent of all properties of the Earth other than its moments of inertia. This means that the motion of the angular momentum vector derived for a rigid Earth may be applied to any other reasonable model.

2. The instantaneous rotation axis of the Earth

A rigorous definition of this axis is valid only in the case of a rigid Earth. The equations of motion of this axis can be obtained by small changes of the coefficients of the periodic terms in the equations governing the motion of the angular momentum vector. In just this manner the ζ-axis of the terrestrial ephemeris has been defined in textbooks and astronomical ephemerides.

3. The Jeffreys-Atkinson axis

The gravitational torques exerted by the Moon and the Sun on the Earth's equatorial bulge not only force the angular momentum vector of the Earth to change its orientation in space, but they also cause a small departure of the axis of figure (i.e. the axis of maximum inertia) from the unit vector $\overline{h}$. The action of each of these bodies can be treated separately.

Let $\overline{m}$ be the unit vector in the direction of the Moon or Sun, and $\overline{f}$ be the unit vector along the axis of figure of the Earth. The vectors $\overline{m}$, $\overline{h}$, and $\overline{f}$ can be shown to lie approximately in the same plane, whose position is defined by the known coordinates of the disturbing body (Figure 2). In the case of a rigid Earth the angle between the unit vectors $\overline{h}$ and $\overline{f}$ can be computed for any moment of time. This enables the direction of $\overline{f}$ to be predicted and used as the ζ-axis of the terrestrial reference system. This idea was first suggested by Jeffreys (1963), and then elaborated by Atkinson (1973, 1975). Murray (1978) thinks that the same ephemeris axis should if possible be retained for other Earth models.

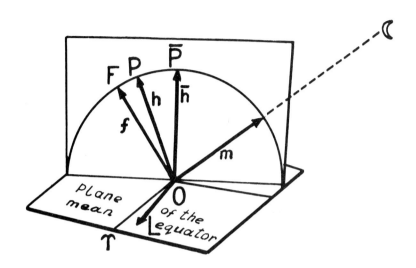

Figure 2. Relationship among the axis of figure, angular momentum vector, and the direction to the Moon or Sun.

The benefit of such a choice is that the unit vector $\bar{f}$ does not move rapidly with respect either to the xyz system rotating with the Earth or to a non-rotating frame attached to remote sources. The IAU General Assembly in Grenoble, 1976, recommended that the ephemeris axis, ζ, should be redefined in the manner proposed by Atkinson but avoided using the term "axis of figure". This was done because the motion of this axis is known to consist of free and forced components while the ζ-axis defined in accordance with Jeffreys' and Atkinson's suggestions is affected by only the forced motion and does not coincide with the axis of figure. Atkinson (1975) sometimes uses the term "axis of figure" in referring to the forced motion of the axis of figure and at other times to refer to the conventional z-axis. He writes, "We now define as the 'pole of figure' the adopted mean pole from which meridian observers reckon their colatitudes ... any constant adopted colatitudes will be adequate, assuming that they are roughly correct." It may be due to this lack of consistency in terminology that the following resolution was adopted in Kiev in 1977:

> IAU Symposium No. 78 recommends that the decision of the sixteenth General Assembly of the IAU that "the tabular nutation shall include the forced periodic terms listed by Woolard for the axis of figure" shall be annulled and that the nutation of

the true pole of date with respect to the mean pole of date should be computed for the motion of the instantaneous axis of rotation.

It has been mentioned that the direction of the angular momentum vector, $\overline{H}$, (or of the unit vector, $\overline{h}$) is the same for the rigid, elastic, or any other reasonable Earth model. However, the relative motion of the vectors $\overline{h}$ and $\overline{f}$ substantially depends on the mechanical properties of the Earth. According to McClure (1973) the effect of tidal deformation of the elastic Earth manifests itself in the relative motion of these vectors with an amplitude reaching two seconds of arc.

It has been pointed out already that the Jeffreys-Atkinson axis can be defined as an axis that would have no short-period motion either in space or relative to the terrestrial frame. This conceptual definition was first applied to the rigid Earth, but it can be extended to elastic models with a liquid core. The problem now is to replace the rigid Earth in the theory of precession and nutation with another model better fitted to our current knowledge of the mechanical properties of the Earth and to derive for this model the equations of motion of the axis which satisfy the requirement that it should only change its direction slowly on the time scale of a day with respect to both the celestial and terrestrial frames of reference. It is to be expected that the equations of transition from the vector $\overline{h}$ to this axis be somewhat different from Woolard's (54). To monitor the motion of the ephemeris ζ-axis with respect to the conventional xyz-system special observations are conducted.

We may pass to the more familiar geometrical representation by constructing the auxiliary unit sphere of Figure 3.

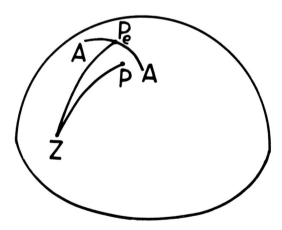

Figure 3. The auxiliary unit sphere.

The point at which the ζ-axis passes through this sphere is the instantaneous ephemeris pole, P. Any conventional system in which the motion of this pole is monitored is linked to the unit vectors $\bar{e}_j$. To define the orientation of the z-axis of this system it is sufficient to adopt two angles which this axis forms with each unit vector $\bar{e}_j$. If we use more than two vectors the values of the angles cannot be prescribed arbitrarily. These angles should be derived from observations.

THE CONVENTIONAL INTERNATIONAL ORIGIN

The IPMS uses the conventional axis attached to the plumb lines at five points on the Earth's surface. The unit vectors, $\bar{e}_j$, parallel to these plumb lines define the zeniths, $\bar{z}_j$, on the auxiliary sphere. Let us take one of the latitude stations and assume that its longitude is known. Then, deriving from an observation at time, t, the instantaneous latitude, ϕ, of the station, we obtain the position of the zenith in the ephemeris system $\xi \eta \zeta$ measuring from the pole P the arc PZ along the meridian of the station equal to its colatitude, $90° - \phi$. By definition the CIO is the point P_e located at the angular distance, $ZP_e = 90° - \Phi$, from the zenith Z, where Φ is the adopted constant value of the mean latitude of the station. In other words, the locus of the conventional pole P_e is a circle, AA described on the auxiliary sphere with Z as the center and spherical radius, $90° - \Phi$.

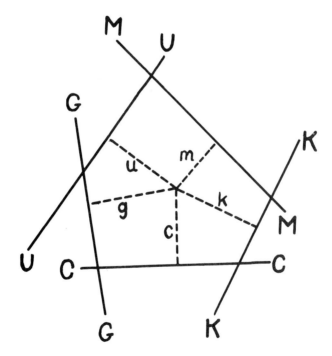

Figure 4. Lines of position defining the CIO.

Considering a small region on the sphere in the vicinity of P (Figure 4) we may replace the arc AA by a straight line located at a distance $(90° - \phi) - (90° - \Phi) = \Phi - \phi$ from the instantaneous ephemeris pole, P, and called the line of position. The CIO is believed to be the point at which all the lines of position of the five international stations always cross. However, from inspection of the observational data we can satisfy ourselves that such a point does not exist at all. Thus, the CIO cannot be defined as the point situated at the constant defining angular distances from the zeniths of the five International Latitude Service stations. The very definition of the CIO should be changed.

It is easy to show that the CIO is the point for which the following always exists:

$$m^2 + k^2 + c^2 + g^2 + u^2 = \text{minimum}. \tag{5}$$

In this condition m, k, c, g, and u represent the distances from the lines of positions of the stations Mizusawa, Kitab, Carloforte, Gaithersburg, and Ukiah respectively.

THE POLE OF THE BIH 1968 SYSTEM

Time observations are also capable of giving the position of the conventional pole, P_e, relative to the ephemeris pole, P. Taking two stations we can obtain the line of position, BB, (Figure 5) such that any point on the line assumed for P_e will preserve the longitude difference of the selected stations. If the zeniths of the stations were fixed to one another all such lines of position will cross at a single point which may be taken for the conventional pole, P_e. This is not the case, and we have to determine the position of P_e by means of a condition similar to (5).

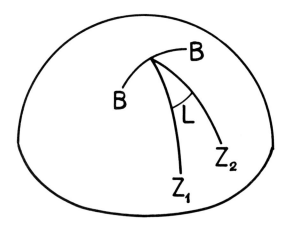

Figure 5. Line of position for the BIH 1968 System.

COORDINATE SYSTEMS AND POLAR MOTION

The realization of the terrestrial reference system adopted by the BIH is achieved by assigning mean longitudes and latitudes to a number of stations. This means that the z-axis is related to the lines of position derived from both time and latitude observations. However, it is not fixed with respect to the unit vectors $\bar{e}_j$ since they do not maintain their directions relative to one another.

THE MEAN POLE OF THE EPOCH OF OBSERVATION (ORLOV'S POLE)

The fact that sum (5) nearly always differs significantly from zero is considered an indication of a nonpolar component in the latitude variations. To separate this component Orlov (1941) compared observations at observatories with nearly the same (or differing by 180°) longitudes. Periodic variations of latitudes proved to be nearly identical, while variations of the mean latitudes (obtained by filtering out periodic components) were quite different.

This has been confirmed by Mironov (1974) who obtained correlation coefficients for a number of pairs of stations with nearly equal longitudes. For periodic variations the correlation coefficients proved to be always positive and only in rare cases smaller than 0.75. On the other hand, divergent values ranging from -0.90 to +0.90 have been obtained for the non-periodic variations. That is a forceful argument in favor of the opinion that variations of the mean latitude are of a non-polar origin and that these variations should be excluded from observations prior to using them for the computation of polar motion. Then, proceeding in the same way as in the case of the determination of the CIO, we shall arrive at the mean pole of the epoch of observation, P_0. The BIH used this pole from 1959 to 1967.

The relative displacement of the CIO and P_0 is called the secular polar motion. Several authors have tried to determine the rate and direction of this motion. The agreement of their results is easily explained since all of them have applied the same methods to the same initial data from the ILS Stations. To estimate the reliability of the results obtained we have derived linear trends from observations from 1900 to 1972 (Fedorov, 1975). The following centennial rates have been obtained:

Mizusawa	$-0\overset{"}{.}330$,
Carloforte	+0.049,
Gaithersburg	+0.241,
Ukiah	+0.345.

Observations at Kitab commenced in 1931 and have been found to be useless for our discussion.

The following null hypothesis has been considered: the observed linear trends are independent random values. Using known methods of statistical testing we have found that the probability of this hypothesis is equal to 0.38 which means the hypothesis does not contradict observations at these four international stations. These data are too scanty for a definite answer as to whether or not the secular motion of the pole has taken place during the last century.

These considerations lead me to believe that the motion of the ephemeris pole should be related to the mean pole of epoch rather than to the CIO. In other words, the z-axis of the conventional terrestrial system should be directed toward P_0. Now we shall consider the general principles of the observations from which the polar motion can be derived.

CLASSICAL ASTRONOMICAL METHODS

Let $\bar{e}$ be the unit vector parallel to the vertical defined in the conventional terrestrial system by its coordinates. The matrix $M(t)$ is not known in advance. Thus, to convert to the non-rotating frame XYZ we must use the "ephemeris" matrix M_0. If $\bar{s}$ is a unit vector directed towards the observed star we may obtain the "ephemeris" or predicted cosine of the angle between $\bar{s}$ and $\bar{e}$ by a scalar multiplication of $M_0\bar{e}$ by $\bar{s}$:

$$\cos \gamma_0 = \bar{s} \cdot M_0 \bar{e}.$$

The observed value of the cosine is

$$\cos \gamma = \bar{s} \cdot (M\bar{e} + \bar{\Sigma}),$$

where $\bar{\Sigma}$ is the sum of errors independent of the relative orientation of the coordinate systems. Therefore

$$\cos \gamma - \cos \gamma_0 = \bar{s} \cdot (M - M_0)\bar{e} + \bar{s} \cdot \bar{\Sigma} = \bar{s} \cdot M_0 \sigma \bar{e} + \bar{s} \cdot \bar{\Sigma}. \tag{6}$$

From (6) one can obtain equations for deriving coordinates of the pole from astronomical observations as well as the difference between universal and atomic time.

RADIO INTERFEROMETRY

The conventional terrestrial system may also be attached to the unit vector, $\bar{e}$, of the baseline of a radio interferometer. Zhongolovitch (1976) emphasizes that such a system will be based on a much more rigid foundation than any using the directions of the plumb lines of several observatories.

The method of observation is based on the same equation (6). If D is the length of the baseline connecting the two antennas and c is the velocity of light, then we immediately obtain from (6)

$$\tau - \tau_0 = s \cdot \frac{D}{c} M_0 \sigma e + s \cdot \Sigma, \tag{7}$$

COORDINATE SYSTEMS AND POLAR MOTION

where $\tau - \tau_0$ is the difference in the time of arrival of the signal at the two antennas.

LASER RANGING TO SATELLITES AND REFLECTORS ON THE MOON

In this case $\bar{e}$ refers to the unit vector along the geocentric position vector of the observing station. The construction shown in Figure 6 is made in the object space, but the only information required for Earth rotation is the variation in the direction of $\bar{e}$ (or the attached xyz system) in the image space.

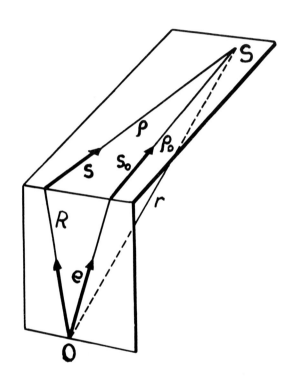

Figure 6. Geometry used in the determination of Earth rotation by laser ranging.

We denote by ρ the distance from the station to the satellite and by R the modulus of the position vector of the station. Then we may write the basic equation,

$$\rho \bar{s} = \bar{r} - RM\bar{e} + \bar{\Sigma}. \qquad (8)$$

For the "ephemeris" value of the "station-satellite" vector we have

$$\rho_0 s_0 = r_0 - RM_0 \bar{e}. \qquad (9)$$

Hence

$$\rho\bar{s} - \rho_0\bar{s}_0 = \rho(\bar{s} - \bar{s}_0) + (\rho - \rho_0)\bar{s}_0 = -RM_0\sigma e + \bar{\Sigma}, \qquad (10)$$

where the difference $r - r_0$ is included in $\bar{\Sigma}$. Scalar multiplication of (10) by s_0 taking into account that $s \cdot (s - s_0) = 0$ leads to

$$\rho - \rho_0 = -\bar{s}_0 \cdot RM_0\sigma e + s_0\bar{\Sigma}. \qquad (11)$$

This equation underlies the methods of deriving the coordinates of the pole from laser ranging results. Scalar multiplication of (10) by a vector, $\bar{n}$, normal to $\bar{s}$ leads to the expression,

$$n \cdot (\bar{s} - \bar{s}_0) = -\frac{R}{\rho}\bar{n} \cdot M_0\sigma e + \frac{\bar{n}}{\rho} \cdot \bar{\Sigma}. \qquad (12)$$

The dynamical method by which polar coordinates can be determined from observations of the directions to satellites are based on the general expression (12).

It should be realized that the coordinates of the pole enter equations (7), (11), and (12) through the matrix σ. Since the right-hand parts of these equations contain the matrix M_0 quasi-diurnal variations should be seen in the differences $\rho - \rho_0$, $\tau - \tau_0$, and $\bar{s} - \bar{s}_0$.

Depending on the methods and techniques employed, the axes of the conventional terrestrial system are related to different unit vectors. Comparison of the polar coordinates obtained in different systems is capable of giving information on the relative motion of the z-axis of these systems since the position of the ephemeris pole in space is independent of the choice of the conventional terrestrial reference frame.

REFERENCES

Atkinson, R. d'E.: 1973, Astron. J. 78, p. 147.
Atkinson, R. d'E.: 1975, Mon. Notices Roy. Astron. Soc. 171, p. 381.
Brandt, W. E.: 1975, Astron. Zu. Moskva 52, p. 1096.
Clemence, G. M.: 1963, "Stars and Stellar Systems" vol. 3, p. 1.
Fedorov, E. P., Korsun, A. A., and Mironov, N. T.: 1972, in P. Melchior and S. Yumi (eds.), "Rotation of the Earth", D. Reidel Co., Dordrecht, p. 78.
Fedorov, E. P.: 1975, Astrometrija i Astrofizika, Kiev, No. 27, p. 3.
Fedorov, E. P.: 1976a, Astrometrija i Astrofizika, Kiev, No. 29, p. 18.
Fedorov, E. P.: 1976b, Nabludenija Iskustvennykh Nebesnykh Tel., No. 15, Moskva, p. 253.
Jeffreys, H.: 1963, Foreward to E. P. Fedorov, "Nutation and Forced Motion of the Earth's Pole", Pergamon Press, London.
McClure, P.: 1973, "Diurnal Polar Motion", Goddard Space Flight Center Report X-592-73-259.
Mironov, N. T. and Korsun, A. A.: in "Proc. Second Int. Symposium of Geodesy and Physics of the Earth", Potsdam, p. 173.

Murray, C. A.: 1978, in press.
Orlov, A. J.: 1941, Bull. Gos. Inst. Sternberg, Moskva, No. 8.
Veis, G.: 1976, in "Proc. IAU Colloquium No. 26", p. 261.
Zhongolovitch, I. D.: 1976, in "Proc. IAU Colloquium No. 26", p. 293.
Zhongolovitch, I. D.: 1977, Trudy Inst. Teor. Astron. Leningrad, No. 16, p. 19.

DERIVATION OF POLE COORDINATES IN A UNIFORM SYSTEM FROM THE PAST ILS DATA

S. Yumi, K. Yokoyama and H. Ishii
International Latitude Observatory
Mizusawa, Iwate
023 Japan

ABSTRACT

The recomputation of the past ILS observations has been carried out at Mizusawa and partly at Cagliari. Preliminary reduction of the observations at the northern stations has been completed. The coordinates of the pole were calculated preliminarily and were compared with those by Vicente and Yumi (1969, 1970). It is known that the coordinates of the pole in the past ILS reports require considerable corrections which are probably mainly due to errors of the micrometer values. Magnetic tapes of the original observational records and of the individual latitudes are now available on request.

1. DATA

Transformation to machine readable form (cards and magnetic tapes) of the original records of the past ILS observations on the northern parallel of 39°8' has been completed. The stations and the periods are: MIZUSAWA (1899-present), KITAB (1930-present), CARLOFORTE (1899-1943, 1946-present), GAITHERSBURG (1899-1914, 1932-present), UKIAH (1899-present), CINCINNATI (1899-1915) and TSCHARDJUI (1899-1919). Meteorological records and readings of the levels and the micrometer are stored, together with other various notes, on magnetic tapes in a format convenient for reduction.

2. PRELIMINARY REDUCTION

Positions and motions of the stars are taken from the catalog by Melchior and Dejaiffe (1969). Apparent places of the stars are computed after the method by Yumi, et al. (1974). The nutation is based on Woolard's (1953) table for the rotation axis. Instrumental constants other than the micrometer value (level constants, inequality of the micrometer, etc.) are based on those in the past reports of the ILS, since these quantities can not be reproduced anymore, nor can we derive corrections to them from the observations. Provisional micrometer values are determined from the past ILS reports taking into account their drastic changes. Corrections to the provisional micrometer values are determined by a new method which is different from the traditional one.

3. ERROR SOURCES

An observed latitude reduced through the procedure mentioned above is considered to include various kinds of errors. The following three are the most important error sources: i) errors of the micrometer value, ii) declination and proper motion errors, and iii) errors in the nutation. Since the three quantities above are closely interrelated, separation of them from one another is one of the most troublesome problems in the course of the reduction. There are two points to be resolved for determining the above three quantities independently.

First, the absolute corrections to the provisional values of the micrometer should be determined instead of the traditional relative ones. The relative correction is expressed as $dM^s - \overline{dM}$, where dM^s and $\overline{dM}$ denote the absolute correction to the provisional micrometer value of the station s and the mean of dM^s averaged over all the stations, respectively. Individual latitudes corrected for the relative micrometer corrections, are affected by $\overline{dM}$ multiplied by the observed zenith distances, which gives rise to spurious non-polar latitute variation common to all the stations. This has made a precise determination of the nutation amplitudes difficult. Furthermore, for the purpose of eliminating the effect of $\overline{dM}$ in the declination and the proper motion errors, the method of successive approximation has been traditionally adopted. Adoption of our new method makes it possible to omit this complicated process.

Second, since one of the important purposes of the project of the recomputation of the past ILS data is in the precise determination of the nutation amplitudes, a new method should be devised for separating the nutation errors from the errors of the star places. If the nutation errors are not taken into account in application of the chain method, derived group corrections are affected by the nutation errors. For example, the semiannual nutation error is absorbed in the position correction and the principal nutation error is absorbed in the proper motion correction if the data span does not cover 18.6 years. Recently, Manabe, et al. (1978) devised a new method for determining corrections to the star places together with the corrections to the nutation amplitudes of arbitrary frequencies. The results of the test calculation with this method proved that both the nutation amplitudes and the corrections to the star places could be determined simultaneously with satisfactory accuracies. We are now intending to extend this method to derive corrections to the star places of all the programme star pairs since 1899 in a uniform system by solving a set of equations with more than one thousand unknowns (more than 500 star pairs and two unknowns for each, that is, position and proper motion corrections).

4. DETERMINATION OF ABSOLUTE MICROMETER CORRECTIONS

Let us explain the method briefly, since details are given in the Vol. XII of the ILS results (Yumi and Yokoyama 1978).

An observed latitude of the station s at the epoch t, $\phi^s(t)$, is expressed as,

$$\phi^s(t) = \phi_p^s(t) - d\delta(t) - D(t) \cdot dM^s(t), \qquad (1)$$

$\phi_p^s(t)$: true latitude including the non-polar common variation,
$d\delta(t)$: correction to the adopted declination,
$D(t)$: observed zenith distance in micrometer turns, which is considered to be common to all the stations for each pair,
$dM^s(t)$: correction to the provisional micrometer value per ½ turn.

Each of the above quantities is considered to be a monthly mean in the following discussions. Let us solve, by the method of least squares, the following equation with two unknowns, $A^s(t)$ and $B^s(t)$, for each of the two or three groups every month,

$$\phi^s(t) = A^s(t) + \gamma \cdot D(t) \cdot B^s(t) \qquad (2)$$

where γ is a normalizing factor to be taken into account if the provisional micrometer value is far from 40"/turn. The number of equations (2), equivalent to the number of pairs in a group is six or eight. One of the solutions $A^s(t)$ has the same meaning as the control latitude (Markowitz 1961), but in a slightly modified way. We deduce the absolute micrometer correction from the solution $B^s(t)$. The quantity $B^s(t)$ can be expressed as

$$B^s(t) = -dM^s(t) - \varepsilon(t). \qquad (3)$$

$\varepsilon(t)$ is a quantity peculiar to each group, since it depends only on $d\delta(t)$ and $D(t)$ of the star pairs in each group. Therefore, $\varepsilon(t)$ can be estimated by the ordinary chain method on the assumption that $dM^s(t)$ does not vary drastically among the evening, intermediate and morning series. The chain method gives the solution in a form,

$$E(t) = \varepsilon(t) - \overline{\varepsilon}(t), \qquad (4)$$

where $\overline{\varepsilon}(t)$ is considered to be the mean of $\varepsilon(t)$ of all of the twelve groups, which is caused inevitably by an application of the chain method. $\overline{\varepsilon}(t)$ is an unknown at present. Combining equations (3) and (4), we have

$$'B^s(t) = B^s(t) + E(t) = -dM^s(t)/\gamma - \overline{\varepsilon}(t). \qquad (5)$$

The observed latitude of equation (1) corrected for $'B^s(t)$ is written as

$$'\phi^s(t) = \phi_p^s(t) - d\delta(t) + D(t) \cdot \overline{\varepsilon}(t) \cdot \gamma. \qquad (6)$$

For estimating $\bar{\varepsilon}(t)$, we use the reduction to group mean. Assuming that $d\delta(t)$ is not correlated with $D(t)$, we can neglect $d\delta(t)$ included in the reduction to group mean and can infer $\bar{\varepsilon}(t)$. Thus, the effects of the errors of the star places could be eliminated from $B^S(t)$ of equation (3). Through this procedure, the absolute corrections to the provisional micrometer values can be determined independently for each station. The final values of $E(t)$ and $\varepsilon(t)$ which are principally independent of localities are determined from $B^S(t)$ of all the stations. Actually, $E(t)$ and $\varepsilon(t)$ derived for each of the ILS stations independently, agree very well among all the stations. The only exception during all the ILS observations, which gives values of $E(t)$ and $\bar{\varepsilon}(t)$ divergent from the other stations is CINCINNATI from 1912 through 1922. In this period, estimated corrections of the CINCINNATI micrometer show very unstable variations.

Most of the temperature coefficients of the micrometers adopted in the past ILS reports require considerable corrections. Consequently, the past reductions may have yielded spurious annual variation of latitude. The individual latitudes are corrected by a linear interpolation of $dM^S(t)$, first taking into account the corrections to the temperature coefficients.

5. DETERMINATION OF THE CORRECTIONS TO THE STAR PLACES

In order to test whether or not the correction to the micrometer values was done well, corrections to the star places were derived following the traditional method using the reduction to group mean and the group correction, based on the individual latitudes and taking into account the absolute micrometer corrections. Derived results agree generally very well among all the ILS stations. The point to be emphasized from this fact is that the micrometer correction of our method was done quite accurately. If the micrometer correction is not accurate, the correction to proper motion, $d\mu'$, especially will deviate remarkably, since the zenith distances become large due to precession at the beginning and the end of the observing program.

6. PRELIMINARY DERIVATION OF THE COORDINATES OF THE POLE

After applying the reduction to group mean and the group correction, the monthly mean latitudes were calculated, assigning weights by the number of observations. In order to compare with the coordinates of the pole given in the past ILS reports, the preliminary values of (x, y) were calculated with the same combinations of the stations as used in the past calculations. Therefore, (x, y) from 1922.7 through 1934 were derived with the three stations of MIZUSAWA, CARLOFORTE and UKIAH, and the data of CARLOFORTE from 1941 through 1943 were discarded as was done in Vol. IX. The mean latitudes of CINCINNATI and TSCHARDJUI are taken from the Band V. Weights are not assigned for computing (x, y). Although the system of the Melchior-Dejaiffe catalog is different from that of the GC catalog on which the mean latitudes defining the CIO are based, adoption of the presently used mean latitudes does not change the reference system from the CIO.

7. COMPARISON WITH THE PAST RESULTS AND EXPECTED REVISION OF THE POLAR MOTION

In order to estimate how much of the polar motion in the past ILS reports will be revised, our preliminary results were compared with (x, y) by Vicente and Yumi (1969, 1970), the origin of which is the CIO. Values of (x, y) of the present calculation show remarkable differences from those by Vicente and Yumi. Behavior of the differences consists of both irregular variation and an annual one, the amplitude of which sometimes attains 0''.04. These differences change their appearance drastically, corresponding to changes of the observing program. It is reasonable to attribute these differences to insufficient micrometer corrections in the past reports, especially for the periods before 1922.7 and from 1935 through 1948 when the micrometer corrections were not applied month by month. Differences in the x component from 1949 through 1954, and those in the y component from 1922.7 through 1934 show remarkable annual variation with an amplitude of about 0''.03, although the micrometer corrections were determined very carefully. However, the temperature coefficients during these periods adopted in the past calculations differ considerably from our present determinations. This is the most probable cause for explaining these annual variations in the differences. It is well known that the annual component of the polar motion during these periods shows unusual behavior when compared with other periods. It is thus expected that the annual component of the polar motion will be revised remarkably by the results of the recomputation of the ILS observations, due mainly to the detailed correction of the micrometer values.

Furthermore, there seems to be a sudden change of the system of the polar coordinates around 1918, when CINCINNATI and TSCHARDJUI stopped observations (1915 and 1919, respectively). This may suggest that the mean latitude of these stations require redetermination for defining them in the CIO system.

8. SEPARATION OF NUTATION AND STAR PLACES

Declination and proper motion corrections of all the star pairs observed since 1899 should be determined in a unique system, independent from the nutation. In other words, simultaneous determination of the star places and the nutation is preferable. A test calculation was made for the data during 1935-1954, 1955-1966 and 1967-1977, respectively. The main purpose of the test calculation is to confirm whether the theory by Manabe, et al. (1978) is applicable to the ILS data or not. The semiannual and fortnightly nutation amplitudes were determined with sufficient accuracies from the three group observations, and coincidence with other observational determinations is very good. The principal nutation (with a 18.6-year period), on the other hand, can be determined only from the data during 1935-1954, and the reasults obtained are (Sakai, 1978):

$$\Delta\varepsilon = 9\overset{"}{.}199 \pm 0\overset{"}{.}0035, \qquad \sin \varepsilon\Delta\psi = 6\overset{"}{.}836 \pm 0\overset{"}{.}0035.$$

These results may be the most reliable ones at the present stage for the principal nutation, since the data used for this analysis are reduced in a uniform system, the micrometer correction was done absolutely, and the duration of observations covers twenty years. The semiannual nutation can not be determined from the two group observations, because the right ascension covers only four hours each month and it is difficult to separate the nutation error from the position errors of the declination. In the future, we intend to add the further unknowns of the deflection of the vertical due to the luni-solar tidal forces.

9. MAGNETIC TAPES AVAILABLE ON REQUEST

The following magnetic tapes i), ii), iii) are available and magnetic tape iv) is in preparation.
i) Original records of observations with the meteorological data and the level and micrometer readings,
ii) Individual latitudes based on the provisional micrometer values,
iii) Individual latitudes based on the absolute micrometer values determined by the method in section 4 in this paper,
iv) Magnetic tapes iii) corrected for declination and proper motion corrections in the uniform system (now in preparation).

The definitive values of the coordinates of the pole since 1899 based on the recomputation will be published at the XVIIth General Assembly of the IAU, after resolving various problems stated in this paper.

REFERENCES

Kinoshita, H.: 1977, Celes. Mech. 15, p. 277.
Manabe, S., Sasao, T., and Sakai, S.: 1978, Publ. Int. Latit. Obs., Mizusawa, Vol. XI, No. 2, p. 23.
Markowitz, Wm.: 1961, Bull. Geodes. 59, p. 29.
Sakai, S.: 1977, Proc. Japanese Symposium on the Rotation of the Earth, held in January 1978, at the Hydrographic Institute of Japan.
Vicente, R. O., and Yumi, S.: 1969, Publ. Int. Latit. Obs., Mizusawa, Vol. VII, No. 1, p. 41.
Vicente, R. O., and Yumi, S.: 1970, Publ. Int. Latit. Obs., Mizusawa, Vol. VII, No. 2, p. 109.
Woolard, E. W.: 1953, Astron. Papers of the American Ephemeris and Nautical Almanac, Vol. XV, Part I, Washington.
Yumi, S., Hurukawa, K., and Hirayama, T.: 1974, Publ. Int. Latit. Obs., Mizusawa, Vol. IX, No. 2, p. 175.
Yumi, S., and Yokoyama, K.: 1978, Results of the International Latitude Service, Vol. XII, for 1955 - 1966, Mizusawa.

ON THE COMPUTATION OF ACCURATE EARTH ROTATION
BY THE CLASSICAL ASTRONOMICAL METHOD

Martine Feissel
Bureau International de l'Heure
Paris, France

Improved series of Universal Time and latitude measurements back to 1962 have been provided to the BIH by several observatories recently. New techniques are currently or will soon provide Earth rotation data that are independent from the classical astronomical observations. In the meantime, the BIH has acquired experience on possible methods for achieving better accuracy. These reasons make it worthwhile to apply our present practical knowledge to the past data. The method which will be used for computing Earth rotation data from the updated BIH files is presented.

INTRODUCTION

The worldwide net of about eighty stations which provide Earth rotation data by classical astronomical methods may be considered as a unique instrument, the function of which is to monitor the orientation of the Earth in an inertial reference frame. As any measuring instrument, it may be described by its accuracy and precision. Only accuracy will be considered in the present paper. An instrument is accurate when it is able to provide measures without any systematic trend, or, in other words, when its measurements remain fixed to the same reference frame.

In the operation of the global astronomical instrument for measuring Earth rotation, the Earth is represented by a sphere of station zeniths (Danjon, 1962), and the inertial reference frame is provided by stars that are observed through dedicated telescopes. Accuracy can be achieved by a comprehensive analysis and an adequate correction of all spurious effects which may appear. The questions which are to be answered for this purpose are the following:
 1 - What causes deformations to the sphere of zeniths, and how should these effects be minimized;
 2 - What causes motions and deformations to the celestial reference frame, and how should their effects be minimized;
 3 - If one wants to devise a method to get rid of systematic errors what is the error power spectrum associated with 1 and 2.
The method presented below is a possible answer to these questions. It

will be applied to the improved data provided by the observatories. This new reduction will benefit from the comparison with the completely independent Doppler method for polar motion. It is hoped that it will provide a better basis for future comparisons with lunar laser ranging and long base interferometry determinations of the Earth's rotation.

1. THE CAUSES OF INACCURACY

The perturbations with effects larger than $0\rlap{.}''001$ are listed in Tables 2 and 3. Some comments follow.

1.1. Practical realization of the terrestrial reference frame

Global motions. Some terms of the Earth tides have local effects; some cause a real change in the Earth's rotational speed. Continental drift gives motions that are common to subsets of stations; absolute velocities (Solomon and Sleep, 1974) have to be considered.

Local motions. Real local tectonic motions may take place. The instruments and their operation (Niimi, et al., 1976) and atmospheric conditions (Hughes, 1974) may give rise to large-scale errors and to flicker noise (Barnes, 1969).

Table 1 shows the rms differences between series of observations obtained with similar instruments and identical programs. The averaging time is 0.05 years; when the instruments are not located in the same observatory the BIH results were used as an intermediate reference. The figures in Table 1 show also the large differences which may exist in the precision of the results from identical instruments.

Table 1. Rms differences between identical instruments with the same program.

Instruments	Dates	UT	Latitude
2 astrolabes	1971.05 - 1971.40	$0\rlap{.}^s0128$	$0\rlap{.}''097$
2 astrolabes	1971.75 - 1973.80	0.0201	0.138
2 PZTs	1974.00 - 1975.95	0.0051	0.076
2 PZTs	1975.00 - 1976.95	0.0056	0.043
2 transit inst.	1972.00 - 1974.85	0.0206	--

1.2. Practical realization of the celestial reference frame

Fundamental constants. The erroneous conventional precession constant has no effect (Fricke, 1977); the use of an erroneous value for nutation in longitude gives a common effect on observations (Feissel and Guinot, 1976). The aberration constant is now sufficiently well known, but its change in 1968.0 has effects on the time series.

The FK 4 is intended to provide an inertial reference frame. This is realized to a certain precision. Yet, some instruments are devised in such a way that the stars observed (with some exceptions) cannot be taken from the FK 4, and the programs cannot remain unaltered for many years. We have, then, to consider separately the fundamental system and the local systems used by these instruments.

Fundamental catalog. It has been studied by Fricke (1972), Lederle (1978) and others. Independent information on positions and proper motions is given by catalogs of FK 4 stars which were obtained from instruments used for the determination of the Earth's rotation (Pavlov, et al., 1971; Billaud, 1972; Afanas'eva and Gorshkov, 1974; Billaud, et al., 1978).

Local stellar systems. They usually are taken from the General Catalog. Studies of PZT catalogs (Yasuda and Hara, 1964; McCarthy, 1973; Takagi, et al., 1976; Greenwich, 1976) and those for zenith telescopes (at Blagovestchensk, Borowiec, Engelhardt, etc.) show the initial errors in positions and proper motions of such programs as compared to the fundamental catalog. Changes of programs, of positions, or of proper motions of stars are made in order to improve the local system. A side effect of these changes is an alteration of the local reference system.

2. CORRECTION OF INACCURACIES

We use the following notation:
    t - date; T - local sidereal time of observation;
    $[UT0(i)-UTC](t)$ - UT measurement of station i at date t;
    $[\phi(i)](t)$ - latitude measurement of station i at date t;
    L_i, F_i - reference longitude and latitude of station i;
    $[UT1-UTC](t)$, $x(t)$, $y(t)$ - UT and polar coordinates at date t;
    $\xi(t)$, $\eta(t)$ - coordinates of the pole of the catalog.

The classical equations (1) and (2) for deriving the Earth's rotation are relevant only if the data are accurate.

$$[UT0(i)-UTC](t) = (-x(t) \sin L_i + y(t) \cos L_i) \tan F_i + [UT1-UTC](t). \qquad (1)$$

$$[\phi(i)](t) - F_i = x(t) \cos L_i + y(t) \sin L_i. \qquad (2)$$

Accuracy will be achieved by adding to (1) and (2) correcting terms for all causes of inaccuracy. The chosen method of correction will depend on whether the perturbation involved is constant or variable, modeled or not, local or common to all stations.

Table 2. Perturbations with a constant effect ($\geq 0\overset{"}{.}001$).

Form	Cause and amplitude (UT0(i); ϕ(i))	
periodic	Common	Local
period < 0.1y	Earth tides ($0\overset{s}{.}002$; 0) diurnal nutation (0; $0\overset{"}{.}001$-$0\overset{"}{.}006$)	Earth tides* ($0\overset{s}{.}0005 \tan F_i$; $0\overset{"}{.}01 \frac{\sin}{\cos} 2F_i$) diurnal nutation (($0\overset{s}{.}0001$-$0\overset{s}{.}0004$) $\tan F_i$; 0)
period = 0.5y	Earth tide ($0\overset{s}{.}005$; 0)	Earth tide* (0; $0\overset{"}{.}001 \sin 2F_i$) change of aberration constant ($0\overset{s}{.}0001 \sec F_i$; $0\overset{"}{.}001 \sin F_i$)
period = 0.95y	diurnal nutation (0; $0\overset{"}{.}001$)	diurnal nutation ($0\overset{s}{.}0001 \tan F_i$; 0)
period = 1.0y	Earth tide ($0\overset{s}{.}002$; 0) diurnal nutation (0; $0\overset{"}{.}011$) FK 4 position errors function of α($0\overset{s}{.}005$; $0\overset{"}{.}05$)	Earth tide* ($0\overset{s}{.}001 \tan F_i$; $0\overset{"}{.}02 \cos 2F_i$) diurnal nutation ($0\overset{s}{.}001 \tan F_i$; 0) local catalog position errors ($0\overset{s}{.}01$; $0\overset{"}{.}1$) change of aberration constant (0; $0\overset{"}{.}005 \cos F_i$) change of program ($0\overset{s}{.}01$; $0\overset{"}{.}1$) change of star coordinates ($0\overset{s}{.}005$; $0\overset{"}{.}05$)
period = 9.3y	Earth tide ($0\overset{s}{.}001$; 0)	
period = 18.6y	Earth tide ($0\overset{s}{.}15$; 0)	Earth tide* (0; $0\overset{"}{.}001 \sin 2F_i$)
biases	FK 4 position errors function of δ ($0\overset{s}{.}005$; $0\overset{"}{.}05$)	local catalog bias Earth tide* (0; $0\overset{"}{.}01 \sin 2F_i$)
steps		change of aberration constant ($0\overset{s}{.}002 \sec F_i$; 0)
white noise	FK 4 mean error of positions	local catalog mean error of positions
flicker noise	FK 4 mean error of proper motions	local catalog mean error of proper motions

*and deflection of the vertical

Table 3. Perturbations with a variable effect ($\geq 0\overset{"}{.}001/y$).

Form	Cause and amplitude (UT0(i); $\phi(i)$)	
	Common	Local
variation of annual term	nutation (principal term) (0; $0\overset{"}{.}02$, period 18.6y) FK 4: μ errors function of α ($0\overset{s}{.}0001/y$; $0\overset{"}{.}001/y$)	nutation (principal term) ($0\overset{s}{.}002$ tan F_i, period 18.6y; 0) local catalog: μ errors ($0\overset{s}{.}001/y$; $0\overset{"}{.}01/y$) change of star proper motions ($0\overset{s}{.}001/y$; $0\overset{"}{.}01/y$) refraction, instrument ($0\overset{s}{.}002$; $0\overset{"}{.}03$)
drifts	FK 4: μ errors function of δ ($0\overset{s}{.}0001/y$; $0\overset{"}{.}001/y$)	continental drift ($0\overset{s}{.}0001/y$; $0\overset{"}{.}001/y$) instrument, local tectonic motions (extremely variable)
steps		change of program instrument + equipment local tectonic motions
flicker noise		climate observer, plate measurement

The corrections are of three different kinds: 1) conventional expressions for modeled perturbations; 2) addition of auxiliary unknowns

$$w(t)\tan F_i = (\xi(t) \cos T + \eta(t) \sin T) \tan F_i \text{ in (1)},$$

$$z(t) = -\xi(t) \sin T + \eta(t) \cos T \text{ in (2)};$$

3) empirical corrections regularly updated by a prediction method according to the type of noise in the data (Feissel, 1976)

$$C_i(t) = a_i + b_i \sin 2\pi t + c_i \cos 2\pi t + d_i \sin 4\pi t + e_i \cos 4\pi t. \qquad (3)$$

The complete treatment which will be applied is summarized in Table 4.

Table 4. Correction of the perturbation.

Perturbation	Correction
nutation, Earth tides, deflection of the vertical, change of astronomical constants	conventional expressions
pole of catalog (residual motion) proper motions (common error)	auxiliary unknowns $w(t)$, $z(t)$
polar reference $\neq$ CIO star position errors function of δ	initial calibration of L_i, F_i (or a_i)
refraction star position errors function of α	calibration of b_i, c_i, d_i, e_i versus global solution
continental drift, proper motion errors in FK 4, change of program or local catalog	updating of a_i
climatic variations, proper motion errors in local catalog, change of local program or local catalog	updating of b_i, c_i, d_i, e_i
instrumental deformations	optimal prediction of a_i
refraction, observer, thermal effects on instruments	optimal prediction of b_i, c_i, d_i, e_i

REFERENCES

Afanas'eva, P. M., and Gorshkov, V. L.: 1974, Astron. Zh. 51, pp. 652-7.
Barnes, J. A.: 1967, Nat. Bureau of Standards (USA) Report 9284.
Billaud, G.: 1972, Astron. Astrophys. 19, pp. 181-188.
Billaud, G., Guallino, G., and Vigouroux, G.: 1978, Astron. Astrophys. 63, pp. 87-95.
Danjon, A.: 1962, Bull. Astron. 23, pp. 187-230.
Feissel, M.: 1976, 29th Journées Luxembourgeoises de Géodyn. (unpub.).
Feissel, M. and Guinot, B.: 1976, Mitt. Lohrmann Obs. Dresden 33, p. 949.
Fricke, W.: 1972, Ann. Review Astron. Astrophys. 10, pp. 101-128.
Fricke, W.: 1977, Astron. Astrophys. 54, pp. 363-366.
Greenwich Time Report: 1976, January-March.
Hughes, J. A.: 1974, Publ. Obs. Astron. Beograd No. 18, pp. 63-81.
Lederle, T.: 1978, Bull. Centre de Données Stellaires 14, pp. 62-68.
McCarthy, D. D.: 1973, Astron. J. 78, pp. 642-649.
Niimi, Y., Oguma, I. and Matsunami, N.: 1976, Publ. Astron. Soc. Japan 28, p. 693.
Pavlov, N. N., Afanas'eva, P. M. and Staritsyn, G. B.: 1971, Trudy Glavnoj Astron. Observ. V. Pulkovo 78, pp. 4-27.
Solomon, S. C. and Sleep, N. H.: 1974, J. Geophys. Res. 79, p. 2557.
Takagi, S., Murakami, G., Kitago, H., Sakai, S., Iwadate, K.: 1976, Publ. Int. Latitude Obs. Mizusawa 10, pp. 179-191.
Yasuda, H. and Hara, H.: 1964, Ann. Tokyo. Astron. Obs. 8, pp. 162-168.

ON THE RELATIVE MOTION OF THE EARTH'S AXIS OF FIGURE AND
THE POLE OF ROTATION

Angelo Poma and Edoardo Proverbio
Stazione Astronomica Internazionale, Cagliari, Italy and
Istituto di Astronomia dell'Università, Cagliari, Italy

Abstract. The motion of the Earth's axis of inertia has been derived, taking elastic deformation into account, from the polar coordinates determined by the BIH for the period from 1962.0 to 1975.0. Characteristics of the motion of both the pole of inertia and the pole of rotation have been examined. The secular displacement of these poles relative to the pole defined by the low order harmonics C_{21}, S_{21} determined from observations of satellites seems to confirm that the inertial reference axis has an apparent wandering motion within the deformable Earth.

1. INTRODUCTION

The fact that the Earth is a deformable body has been well known for a long time. As a result of several forces of varied nature acting on the Earth's mass (oceans and atmosphere, centrifugal forces, earthquakes, etc.), one must expect that the inertia tensor J is time dependent. Consequently also the Earth's axes of inertia, whose direction cosines are given by the characteristic system

$$(J - \lambda I)(\xi, -\eta, \zeta)^{-1} = 0 \qquad (1)$$

where I is the indentity matrix, cannot be considered as constant with respect to the "fixed" reference frame.

As outlined by Gaposchkin (1968) and Melchior (1972) the position of the Earth's axes of inertia could be determined by very exact artificial satellite observations, since the components of the inertia tensor are related to coefficients of the harmonics of the geopotential.

Unfortunately, as shown in table 1, the observed values of the tesseral low harmonics C_{21}, S_{21}, given by satellite observations are at present incapable of providing definitive information about the motion of the inertial pole since their values are comparable with observational errors. Moreover, we would like to have pole positions every five days in order to compare them with, for instance, the position of the pole of rotation. So we are obliged to try to calculate the ξ and η coordinates of the inertial pole directly from the x and y coordinates of the instantaneous pole of rotation. Attempts to deduce the motion of the inertial pole from the Eulerian equation of motion by taking into account the role of the Earth's deformations have been made by C. Dramba (1964, 1976). The difficulty of obtaining reliable results arises from the need to use accurate astronomical observations and appropriate approximations in the equations of motion.

Table 1. Unnormalized geopotential coefficients and components of inertia tensor ($Ma^2 = 1$)

	GEM 6	SE III	GRIM 2
$10^6 \cdot C_{20}$	-1082.6283	-1082.6370	-1082.6350
$10^6 \cdot C_{21}$	- 0.0012		
$10^6 \cdot S_{21}$	- 0.0041		
$10^6 \cdot C_{22}$	1.5654	1.5362	1.6059
$10^6 \cdot S_{22}$	- 0.8961	- 0.8815	- 0.8807
A	0.329697	0.329699	0.329699
B	0.329703	0.329706	0.329706
C	0.330783	0.330785	0.330785
$10^6 \cdot D$	- 0.0012		
$10^6 \cdot E$	- 0.0041		
$10^6 \cdot F$	- 1.7922	- 1.7630	- 1.7614

2. EQUATIONS OF MOTION

The motion of the free nutation of the deformable Earth about its centre of mass may be described by the Liouville equation (Munk and MacDonald, 1960)

$$\frac{d}{dt}\{(J \cdot \bar{\omega}) + \bar{h}\} + \bar{\omega} \times \{(J \cdot \bar{\omega}) + \bar{h}\} = 0 \qquad (2)$$

where $\bar{\omega}$ is the angular velocity of the reference frame x_i, $\bar{h}$ the relative angular momentum and J the inertia tensor.

Since the axis of instantaneous rotation is close to the axis of figure a perturbation scheme can be used to find approximate solutions of equation (2) (Volterra, 1895, 1898). For this purpose let us consider the right hand reference frame with x_1 axis along the Greenwich meridian, x_2 axis along 90° East and x_3 axis pointing to the

CIO. In this system we put

$$\omega_1 = \Omega x \qquad \omega_2 = -\Omega y \qquad \omega_3 = (1 + m)\Omega$$

Conventionally x and y are the coordinates of the instantaneous rotation pole, Ω is the mean angular velocity of the Earth, 2π radians per sidereal day, and m is the relative change in the length of the day. After neglecting the term of order 10^{-9} and substituting the coordinates of the pole of inertia given by (1), namely,

$$\lambda = C$$
$$\xi = (-D(C-B) + EF)/((C-A)(C-B))$$
$$\eta = (E(C-A) - DF)/((C-A)(C-B))$$

according to the assumed approximation, equation (2) is reduced to the linearized system

$$\frac{\sigma_1}{\Omega} \cdot \frac{dx}{dt} - y + \eta = \varepsilon(\xi-x) \; ; \; \frac{\sigma_2}{\Omega} \cdot \frac{dy}{dt} + x - \xi = \varepsilon(y-\eta) \qquad (3)$$

where $\sigma_1 = A/(C-B)$, $\sigma_2 = B/(C-A)$ and $\varepsilon = F/(C-A) = F/(C-B)$ is a corrective term.

3. COORDINATES OF THE INERTIAL POLE FROM BIH DATA

Equations (3) are the differential equations of the polar motion. If one assumes the functions ξ and η to be known, solutions of (3) can easily be found. Vice-versa we may consider (3) as an algebraic system for the unknowns ξ and η and we can solve it by using the observed values of x and y. Dramba and Stanila (1969), using nearly similar equations, followed this procedure and resolved the systems by the least-squares method assuming σ_1, σ_2 and ε to be also unknown. In our opinion, however, the instability of such a system can cause large errors. On the other hand, parameters σ_1, σ_2 and ε vary more slowly and can be regarded as constants. From table 1 we have derived

$$\sigma_1 = 305.437 \pm 0.010 \quad \sigma_2 = 303.680 \pm 0.032 \quad \varepsilon = -0.00164 \pm 0.00007$$

and we have solved each single equation by means of the iterative method

$$\xi = \xi_o - \varepsilon(y-\eta_o) \; ; \; \eta = \eta_o - \varepsilon(x-\xi_o) \qquad (4)$$

where ξ_o and η_o are the solution of (3) for $\varepsilon = 0$.

By using equations (4) the ξ, η coordinates have been derived from the smoothed x,y coordinates of the rotation pole. The latter were supplied by the BIH every five days for the period 1962.0 - 1975.0. The derivatives dx/dt and dy/dt were computed using the usual five-point Langrangian differentiation formula. The wobbles $P(x - \xi, y - \eta)$ of the instantaneous rotation pole with respect to the instantaneous inertial pole are plotted in Fig. 1. Irregular variations sometimes occur when $x - \xi$ and $y - \eta$ are small. This could result from errors inherent in the method, but could alterna-

Fig. 1. Wobbles of the instantaneous rotation pole with respect to the instantaneous inertial pole.

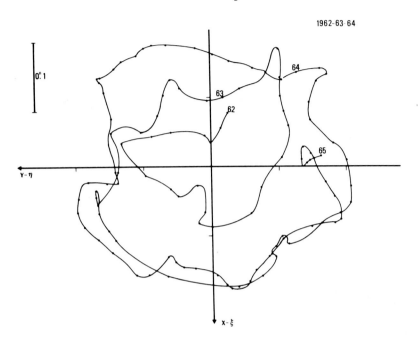

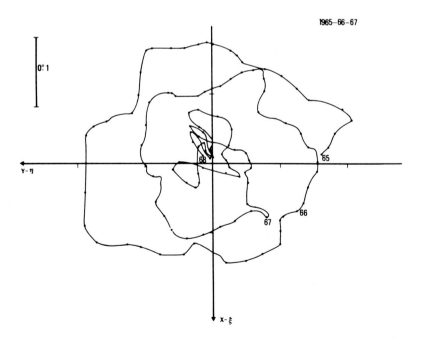

RELATIVE MOTION OF EARTH'S AXIS OF FIGURE AND POLE OF ROTATION

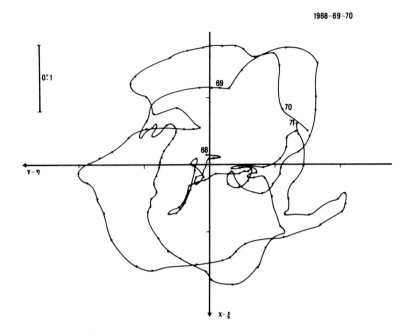

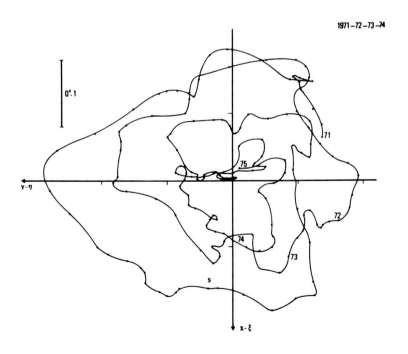

tively be a physical consequence of the fact that $x = \xi$ and $y = \eta$.

Spectral analyses have been carried out on the series $F(t)$ of (x,y), (ξ, η) and $(x - \xi, y - \eta)$ coordinates by means of the conditional equations

$$F(t) = A_a \sin(2\pi t/P_a + F_a) + A_c \sin(2\pi t/B_c + F_c).$$

Two principal periods were emphasized, namely the annual (368 days) and the Chandler period (432 days).

Table 2 Periodical components of the rotation and inertial poles

	P_a	P_c	A_a	F_a	A_c	F_c
x	368^d	432^d	0."103	$209°$	0."126	$346°$
y	368	432	0.091	302	0.133	76
ξ	368	432	0.027	199	0.032	346
η	368	432	0.008	300	0.044	77
x- ξ	368	432	0.074	212	0.094	346
y- η	368	432	0.084	301	0.089	76

The results, given in Table 2, are in good agreement with those found by other authors.

Finally a 6-year running filter was used to derive mean values of the coordinates, free from both annual and Chandler components, at intervals of 1 year; the results are shown in Table 3. It can be seen that both inertia and rotation poles have a similar secular motion. This result seems to confirm the existence of a secular wandering motion of the Earth's rotation axis; but, on the other hand, it could be only an immediate consequence of the equation of motion.

If the observed secular motion of the pole of rotation were really due to the secular drift of the pole of inertia, the results obtained by astronomical observations would be comparable with those derived by satellite observations. However, such a comparison today gives us poor results because, as has been said, few and inaccurate data are generally available. The comparison of the mean BIH pole of inertia for the epoch 1968 with the mean pole derived from C_{21} and S_{21} by GEM 6 (Smith et al., 1976) and GEM 8 (Wagner et al., 1976), given in table 4, shows that the derived secular variations are in very poor agreement.

Table 3. Annual means of the coordinates of the rotation and inertial poles after 6 year running means.

Year	x	y	ξ	η	x-ξ	y-η
1965	-0."0031	0."2379	-0."0029	0."2378	-0."0001	0."0001
1966	- 0014	2364	- 0011	2363	- 0003	0001
1967	0024	2380	0020	2376	0004	0004
1968	0035	2372	0039	2365	- 0004	0007
1969	0064	2399	0071	2393	- 0007	0007
1970	0101	2418	0110	2416	- 0009	0002
1971	0142	2457	0145	2456	- 0003	0001

Table 4

	ξ	η
GEM 6	+ 0."213	- 0."725
GEM 8	+ 0."023	+ 0."069
BIH	+ 0."006	+ 0."240

The values of GEM 6 are one order of magnitude higher than those of GEM 8 while the latter are in but moderate agreement with the data derived from astronomical observations. So only a drastic improvement in the accuracy of satellite observations will confirm the existence or not of a secular trend in the position of the inertial reference axis.

REFERENCES

Balmino, G., Reigber, C., and Moynot, B.: 1976, "Deutsche Geoadatische Kommission" Reihe A, Heft No. 86

Dramba, C.: 1964, "Studii si Cercetari de Astronomie", tome 9, No. 1.

Dramba, C.: 1976, "Rendiconti del Seminario della Facoltà di Scienze dell'Università di Cagliari", Vol XLVI, pp. 273-280.

Dramba, C., and Stanila, G.: 1969, "Studii si Certari de Astronomie" tome 14, No. 1.

Gaposchkin, E.M.: 1968, "Proc. of the Symposium on Modern Questions of Celestial Mechanics", Centro Internazionale Matematico Estivo.

Gaposchkin, E.M.: 1973, "Smithsonian Standard Earth III", SAO Special report No. 353.

Melchior, P.: 1972, in P. Melchior and S. Yumi (eds.), "Rotation of the Earth", IAU Symp. 48, pp. XI-XXII.

Munk, W.H., and MacDonald, G.J.F.: 1960, "The rotation of the Earth", Cambridge Univ. Press, England.

Smith, D.E., Lerch, F.J., Marsh J.G., Wagner, C.A., Kolenkiewicz, R., and Khan, M.A.: 1976, "J. Geophys. Res.", 81, No. 5.

Volterra, V.: 1895, "Atti Accad. Torino", 30, pp. 547-561.

Volterra, V.: 1898, "Acta Math.", 22, pp. 201-357.

Wagner, C., Lerch, F.J., Brownd, J.E., and Richardson, J.A.: 1976, GSFC Report X-921-70-20.

DISCUSSION

J.D. Mulholland: How can you separate the "secular" motion of the pole from secular errors in the orbit of the satellite?

E. Proverbio: We cannot.

SECULAR VARIATION OF TASHKENT ASTRONOMICAL LATITUDE

V. P. Shcheglov and G. M. Kaganovsky
Astronomical Observatory of Tashkent
U.S.S.R.

ABSTRACT

The mean astronomical latitude of the Tashkent Observatory was determined by D. D. Gedeonov in 1895-1896 before the International Latitude Service had been organized. The determination was based on a continuous fourteen-month series of observations with a Wanschaff visual zenith telescope (d = 68 mm., f = 870 mm.). These data consisted of the observations of 2214 Talcott pairs which were observed from July, 1895 through August, 1896.

In accordance with the suggestions and instructions of V. P. Shcheglov, G. M. Kaganovsky repeated the research of Gedeonov in 1969-1970, observing 2369 Talcott pairs in fourteen months. Unfortunately the Wanschaff visual zenith telescope was not preserved, so the observations were carried out with a Bamberg transit instrument (d = 100 mm., f = 1000 mm.). In both cases the observations were reduced to the AGK 4 system.

The variation of the mean astronomical latitude of Tashkent during the 75-year interval between the two observations is (Kaganovsky, 1972)

$$\phi_{1970} - \phi_{1896} = -0\overset{''}{.}309 \pm 0\overset{''}{.}021.$$

The variation of the Tashkent mean latitude in the same time interval derived from the polar coordinates of Vicente and Yumi (1969) is $-0\overset{''}{.}256$. The agreement of these values corroborates secular polar movement. Therefore when geographic coordinates are obtained it is necessary to point out the epoch of their determination and to reduce them to the Conventional International Origin.

REFERENCES

Gedeonov, D. D.: 1899, Astron. Nachr. 148.
Kaganovsky, G. M.: 1972, Doklani Acad. Sci. Uzbekistan S.S.R., no. 7 T.
Vicente, R. and Yumi, S.: 1969, Publ. Int. Latitude Obs. Mizusawa 7, p.1.

DISCUSSION

S. K. Runcorn: Is the secular polar motion that is found in agreement with that found by Markowitz?

V. P. Shcheglov: Yes, because he used the same initial data.

COORDINATES OF THE POLE FOR THE PERIOD 1968 - 1974 COMPUTED
IN THE SYSTEM OF 10 STATIONS WITH SMALL VARIATIONS OF MEAN
LATITUDES

B. Kolaczek
Planetary Geodesy Department
Space Research Center
Polish Academy of Sciences

ABSTRACT

The polar orbit was computed in a system of ten stations with small variations of mean latitudes and compared with the polar orbit based on ten stations with large variations of mean latitudes.

INTRODUCTION

Mean latitudes of stations computed by Orlov's filters or others are changeable. These variations are mainly irregular and secular. The biggest irregular variations are of the order of $0''.1$, but may sometimes reach $0''.2$ and last one to three years (Fedorov et al., 1972; Kolaczek, 1977). The largest secular variations of mean latitudes reach several thousandths of seconds of arc per year such as in the case of Mizusawa (Fedorov et al., 1972). Mean latitude variations can be caused by changes of instruments or observational programs as well as by geophysical phenomena. Mean latitude variations influence polar coordinates and the results of spectral analyses of latitude variations and polar coordinates. They can not be neglected in the study of polar motion with the presently achieved accuracy.

RESULTS AND CONCLUSIONS

In order to show the influence of mean latitude variations the coordinates of the pole for the period 1968 - 1974 were computed according to the IPMS,L formula for the set of stations shown in Table 1. The first set includes the ten stations which have small variations of mean latitudes and mostly high weights in the IPMS and BIH solutions. The second set includes ten stations with large mean latitude variations (on the order of $0''.1$ or larger).

Graphically smoothed IPMS monthly mean latitude data were used for the computations. The results are shown in Figures 1 and 2. In the first case (Figure 1) the polar orbit is similar to the IPMS,L polar orbit and the accuracy of polar coordinate determinations is of the order of 0".001 to 0".002. In the second case (Figure 2) the polar orbit is very deformed and the accuracy of the polar coordinate determinations is ten times lower. This means that variations of the mean latitude of stations have an important influence on the determination of polar coordinates. In this situation it would be worthwile to create a new system, for

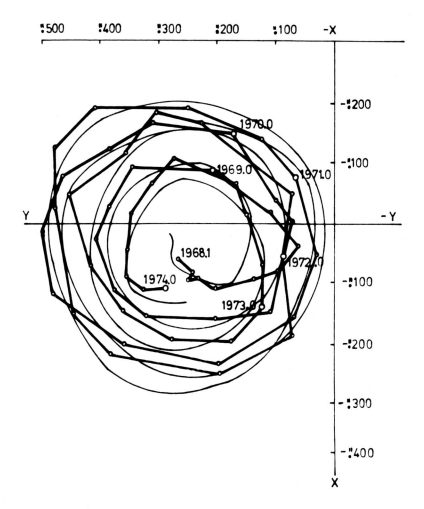

Figure 1. Polar orbit for the first set of observing stations (heavy line). The light line is the IPMS,L polar orbit.

COORDINATES OF THE POLE FOR THE PERIOD 1968-1974

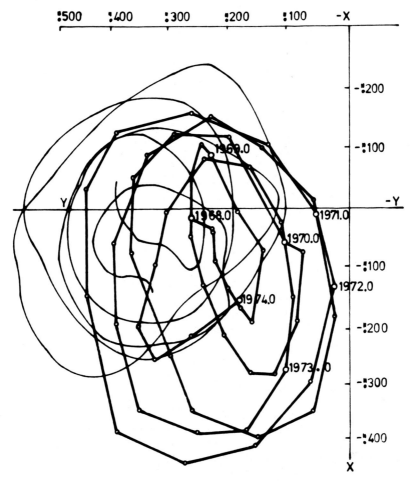

Figure 2. Polar orbit for the second set of observing stations (heavy line). The light line is the ILS polar orbit.

instance IPMS 2, based on some chosen stations having stable instruments and locations in order to study the character of the astronomical systems and the long-period terms of polar motions.

In the case of the ILS stations the Earth's crustal motion has a larger influence on the determined polar coordinates than the star proper motions. Some organization of the observations of all or some of the latitude stations ought to be taken into account to omit too many interruptions of observations due to changes of instruments or observational programs. Nowadays it is difficult to find a period of several years of uninterrupted, homogeneous data for a few stations.

Table 1. Stations used in the study of the effect of large variations of mean latitude on derived polar coordinates.

I		II	
STATION	IPMS WEIGHT	STATION	IPMS WEIGHT
Mt. Stromlo	16	Richmond	5
Blagovestchensk	21	Washington	5
Irkoutsk	16	Gaithersburg	11
Warsaw	11	Mizusawa	3
Pecny	13	Belgrade	31
Greenwich	13	Turku	19
Paris	5	Neuchatel	9
Pulkovo	15	Hamburg	9
Poltava	8	Uccle	8
Tokyo	11	Kitab	9

REFERENCES

Annual Report of the International Polar Motion Service for the years 1962-1974: S. Yumi (ed.), Mizusawa.

Annual Report of the Bureau International de l'Heure for the years 1967-1974: Paris.

Fedorov, E. P., Korsun, A. A., Major, S. P., Panczenko, N. I., Taradij, W. K., Yatskiv, Y. S.: 1972, in "Motion of the Earth's Pole for the Years 1890.0-1969.0" (book ed. by "Naukowa Dumka"), Kiev.

Kolaczek, B.: 1977, "Variations of Differences of Latitudes and of Mean Latitudes of Stations Located in the Vicinity of a Common Meridian", paper presented at IAU Symposium No. 78 - "Nutation and the Earth's Rotation", Kiev.

AMELIORATION DES CALCULS DE REDUCTION DES OBSERVATIONS A L'ASTROLABE.
APPLICATION A LA DETERMINATION DES TERMES DE 18.6 ET 9.3 ANS DE LA
NUTATION

Fernand Chollet
Observatoire de Paris

ABSTRACT

Astrolabe observations made at Paris have been analysed to obtain more
reliable values of the constants of nutation. A new method of
reduction which takes account of variations in the zenith distance of
the observations gave values for the 18.6 year and 9.3 year terms
which are almost the same as those derived by classical methods, but
for the annual and semi-annual terms the results are different. It is
proposed to apply the new method to other series of observations.

INTRODUCTION

Depuis la mise en service des astrolabes de type Danjon, le problème essentiel posé par cet instrument est celui de la stabilité de la distance zénithale d'observation. Divers auteurs ont abordé ce problème de façon théorique (Sheepmaker, 1963.) ou proposé et réalisé des instruments plus stables (Thomas, 1967) et plus précis (Billaud et Guinot, 1971 ; Billaud et Llop, 1975). Mais, comme pour tous les instruments d'astrométrie, les effets de la réfraction sont mal connus et, souvent traités, faute de mieux, de façon peu satisfaisante eu égard à la précision instrumentale.

METHODE PROPOSEE

Il nous est apparu qu'une autre méthode pouvait être envisagée pour réduire, sinon éliminer, l'influence de ces instabilités, si elles existent. Le procédé consiste simplement à tenir compte d'une variation éventuelle de la distance zénithale pendant la durée de l'observation d'un groupe d'étoiles, en prenant ce taux de variation pour inconnue supplémentaire. Cette méthode n'est pas nouvelle puisque divers auteurs l'ont déjà proposée pour d'autres études (Chollet, 1970) ou pour les observations à l'astrolabe (Gubanov, 1975). A notre connaissance, c'est cependant la première fois que la méthode est vraiment mise en oeuvre après quelques résultats préliminaires (Chollet, 1977). Ce procédé présente un avantage supplémentaire car, s'il est efficace, il pourra aisément être appliqué aux séries d'observations obtenues par divers astrolabes en fonction dans le monde depuis 1956. Il n'est d'ailleurs pas interdit de penser qu'il puisse être appliqué à d'autres instruments.

L'introduction de cette inconnue supplémentaire peut avoir pour effet de détruire la précision avec laquelle les inconnues, dites "principales", c'est à dire le temps et la latitude, peuvent être déterminées. Dans ce cas, l'amélioration est illusoire. L'un des buts de ce travail est de montrer qu'il n'en est rien. Les premiers résultats encourageants, obtenus, il a fallu dégager une méthode rapide pour effectuer la re-réduction de quelques 6500 groupes observés à Paris depuis 20 ans.

CALCULS DE REDUCTION

L'utilisation des données d'observations, archivées sur bande magnétique, était possible mais obligeait au calcul des 200 000 positions apparentes des étoiles observées depuis 1956. Nous avons gagné un temps considérable en utilisant les résultats des réductions obtenus par la méthode classique et archivés avec toutes les décimales résultant du calcul. Ces résultats permettaient de reconstituer les équations de conditions à trois inconnues. Il a donc suffit d'ajouter notre quatrième inconnue et de procéder à la résolution par moindres carrés. Aucun calcul de position apparente n'était donc nécessaire. Le gain appréciable sur les temps de calcul ainsi dégagé nous a donné les moyens de régler une question qui n'avait pu être abordée jusqu'à présent. Il s'agit du calcul de nouvelles corrections de lissage interne. On sait que ces corrections ont pour but de rendre comparables les résultats des groupes d'étoiles complètement et incomplètement observés et qu'elles doivent respecter un nombre de relations algébriques égal au nombre des inconnues. Les corrections dont nous disposions ne repondaient plus à ces critères et il était nécessaire de les recalculer. Sur la base des 1000 premières observations, comme pour les réductions "classiques", et en deux itérations, nous avons obtenu un nouveau jeu de corrections de lissage interne, et enfin procédé à la re-réduction de toutes les observations.

RESULTATS

La figure 1 montre, toujours pour les 1000 premiers groupes, l'histogramme des changements de poids classés en fonction des poids eux-mêmes. On constate qu'il y a une amélioration générale des poids. L'augmentation est, en moyenne de 0.67 alors que le poids moyen est de l'ordre de 2.5. Cela correspond à une variation de la dispersion qui passe de 0".20 à 0".18. Le gain est loin d'être négligeable et, en tous cas, du même ordre de grandeur que celui apporté par les améliorations techniques. Il faut noter toutefois que notre but est, non pas d'améliorer la précision interne, mais d'éviter au mieux que certains phénomènes parasites puis-

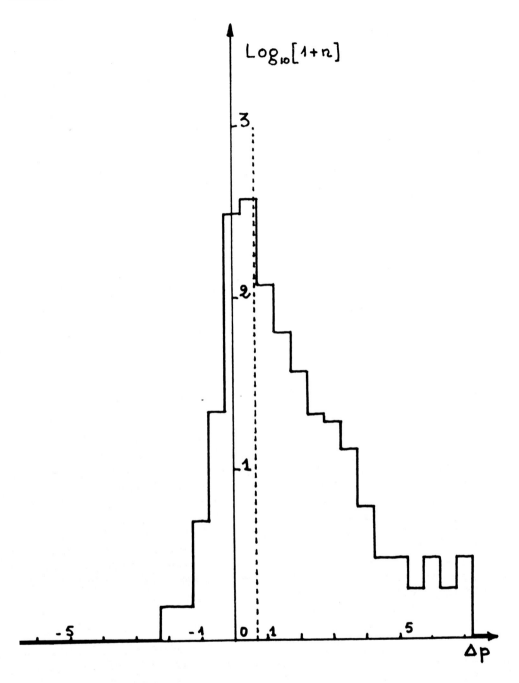

Figure 1: Histogramme des changements de poids, classés en fonction des poids eux-mêmes.

sent fausser les résultats qui nous intéressent. Le fait que la précision interne des observations soit améliorée est un indice de la validité de la méthode et de la réalité de ces effets perturbateurs. On doit encore remarquer que les variations de distance zénithale déterminées par ces calculs, bien qu'obtenues avec une précision plus faible que celle des autres résultats, sont loin d'être négligeables, puisqu'elles atteignent couramment 0".15 à 0".20 par heure. Il nous semble donc d'autant plus intéressant de tenir compte de ces variations de distance zénithale par ce procédé qu'il a l'avantage de compenser, à la fois, les variations instrumentales et les erreurs dans le calcul des variations de la réfraction. Si nous avons quelque peu détaillé les méthodes de calcul, c'est qu'il nous paraît important d'obtenir, avant toute analyse, des résultats aussi fiables que possible. Il apparaît que cette modification dans les méthodes de calcul représente un progrès non négligeable dans la réalisation de cet objectif.

Nous avons donc commencé l'analyse de ces nouveaux résultats et, en particulier, examiné si les constantes principales de la nutation pouvaient être mieux déterminées, et surtout si les corrections nouvelles, déduites de ces observations, changeaient. Pour ce calcul la méthode suivie est, cette fois, tout à fait classique (Guinot, 1958 ; Débarbat et Guinot, 1970 ; Capitaine, 1977) et utilise la méthode des différences de résultats d'observations de groupes d'étoiles observés la même nuit, par le même observateur. Le tableau 1 présente les valeurs des constantes de la nutation déduites de ces nouveaux résultats, ainsi que les valeurs obtenues à partir des résultats classiques. On constate que la précision donnée est du même ordre de grandeur dans les deux cas. Ce fait peut être, en partie, dû à ce que nous utilisons des moyennes annuelles des différences de résultats entre groupes consécutifs. Le fait de procéder à ce lissage tend très sûrement à dégrader la précision des résultats. Il est toutefois plus important de noter que si les termes de 18.6 et 9.3 ans de périodes sont pratiquement identiques dans les deux cas, il en va autrement des termes de périodes annuelles et semi-annuelle, malgré la mauvaise précision de leur détermination. Il semble donc bien que cette méthode de réduction des observations, si elle n'améliore pas la précision finale des résultats, les modifie sensiblement dans certains cas. Il est intéressant de noter que ce sont justement les termes qui ont le plus de chance d'être influencés par l'effet des divers phénomènes géophysiques saisonniers.

CONCLUSION

Il serait donc instructif d'étendre cette méthode de réduction aux données d'observations d'autres astrolabes et peut-être même d'autres instruments. Un autre test que nous avons l'intention d'effectuer est d'évaluer la correction annuelle qu'il faudrait appliquer à ces nouveaux résultats pour les rendre comparables à ceux déduits des données du Bureau International de l'Heure. Enfin, s'il se confirme que cette méthode reste

Tableau 1

	Argu.	Nouvelle Méthode		Méthode Classique	
		Temps	Latitude	Temps	Latitude
$N \sin \varepsilon$	Ω	-6.833 ± 0.011	-6.840 ± 0.007	-6.833 ± 0.009	-6.839 ± 0.006
	2Ω	0.084 10	0.091 7	0.076 9	0.090 6
	$L + \beta$	0.021 77	0.014 16	0.009 70	0.004 15
	$L - \beta$	0.084 77	0.042 22	0.077 70	0.049 20
	$2L$	-0.505 11	-0.506 7	-0.513 11	-0.510 6
	$3L - \beta$	-0.046 23	-0.024 15	-0.048 22	-0.021 13
S	Ω	9.209 ± 0.011	9.216 ± 0.007	9.211 ± 0.009	9.215 ± 0.006
	2Ω	-0.088 11	-0.098 7	-0.095 9	-0.100 6
	$L + \beta$	-0.024 68	-0.009 16	-0.038 63	-0.010 15
	$L - \beta$	0.003 67	0.003 22	-0.006 60	-0.007 19
	$2L$	0.555 11	0.551 7	0.561 11	0.555 6
	$3L - \beta$	0.042 21	0.031 16	0.038 20	0.024 15

applicable, une analyse plus détaillée de ces nouveaux résultats sera entreprise. L'étude des nouvelles corrections de groupes, l'établissement d'un petit catalogue plus précis et enfin l'analyse des variations de la latitude et du temps, sont parmi les objectifs que l'on peut se fixer.

REFERENCES

- Billaud G., Guinot B.1971, Astron. and Astrophys. 11, p. 241.
- Billaud G., Llop H., 1975, Astron.and Astrophys. 41, p. 237.
- Capitaine N., 1977, Communication au Symposium UAI n° 78, Kiev.
- Chollet F., 1970, Astron. and Astrophys. 9, p. 110.
- Chollet F., 1977, Communication au Symposium UAI n° 78, Kiev.
- Débarbat S., Guinot B., 1970 : "La Méthode des Hauteurs Egales en Astronomie". Gordon and Breach.
- Gubanov V.S., 1975 Astron. Zh. 52, p. 857.
- Guinot B., 1958, Thèse Fac. Sci. Univ. Paris. Gautier Villars.
- Sheepmaker A.C., 1963, Publis. Geod. Netherl. Geod. Commis. 1, p. 1.
- Thomas D.V., 1967, Month. Notes Astron. Soc. South Africa, 26, p. 2.

TIME AND LATITUDE PROGRAMS AT THE NATIONAL OBSERVATORY OF BRAZIL

Luiz Muniz Barreto
Observatório Nacional

The National Observatory of Brazil was established in 1827 to study the problems of latitude and time. Time and latitude programs are still an important part of its work, but are now accompanied by increasing activity in the fields of geophysics, astrophysics and radio astronomy.

I do not intend to dwell upon historical aspects of the Time Service of the National Observatory, but will draw to your attention the transmission of electrial time signals between Portugal and Brazil by submarine cable in 1880, and a systematic study of seasonal irregularities in the rotation of the Earth from 1949 to 1955 (1).

Installation of an atomic-clock system began in 1955, and in 1978 this comprises 12 atomic clocks, operating in Rio de Janeiro, São Paulo, Atibaia, Natal and Brasilia. The clocks, separated by distances of up to 1200 km, are intercompared by a TV network (2) supplemented by monthly clock transportation. The two methods maintain a national atomic-time scale with an interval error of less than 0.1 µs.

About 2 comparisons per year have been made with standards from BIH, NBS, USNO and the Observatorio Naval Argentino.

Between 1924 and 1932 Lelio I. Gama made a long series of observations using a visual zenith tube identical with those of the International Latitude Service. This series, comprising 13 000 pairs of observed stars, was published in 1977 (3) and is a unique contribution to this field in the Southern Hemisphere in the first half of this century.

The latitude program has now been improved by the installation of a Danjon impersonal astrolabe. The new program, intended to continue Gama's research, is being carried out in cooperation with the Astronomical and Geophysical Institute of the University of São Paulo.

The astrolabe was installed in January 1977 and a regular program of observations was started in May 1977. We were concerned that our observations might be spoiled by our location within a city of 6 million people, but by taking precautions we have been able to obtain promising results; a special air-extraction system has been arranged in order to reduce local disturbances (4) and the astrolabe has been installed high in the dome to minimise the internal light path.

This arrangement has proved to be excellent for observing the Sun and Venus during the daytime. A planetary positions program is now in progress and regular solar observations to study the position of the equinox are to be started.

The main purpose of the astrolabe is, however, to provide data for the time and latitude programs. Observations of 396 stars in 12 groups have been in progress since May 1977; preliminary analysis shows internal errors of about 7 ms in time and 0.1 arcsec in latitude for each group, with a typical observing error of about 0.3 arcsec for a single star.

The first results will be published in October 1978.

(1) L. Muniz Barreto - 150 Anos de Astromia no Brazil - in preparation.

(2) I. Mourilhe Silva - Comparacão de relógios atômicos por meio de cadeias de televisão - MSc Thesis - ON, 1974.

(3) L. I Gama - Variação de Latitude do Rio de Janeiro (1924.3 - 1931.3) - ON , 1977.

(4) V. A. d'Avila, A. H. Andrei, J. L. Penna, and M. Queiroz, - Cuidados com a refração de sala na instalação do astrolábio Danjon no Observatório Nacional - XXIX Annual Meeting from SBPC, 29, pp 500-501, 1977.

PRELIMINARY ANALYSIS OF ASTROLABE OBSERVATIONS AT MERATE
OBSERVATORY DURING THE PERIOD 1970 - 1977

L. Buffoni, F. Carta, F. Chlistovsky, A. Manara, F. Mazzoleni
Observatorio Astronomico di Milano-Merate
Italia

ABSTRACT

Results of observations made with the Danjon astrolabe at Merate Observatory in the period from 1970 through 1976 were analyzed. The observational program and methods have been discussed previously (Buffoni et al., 1975a, b). The observational accuracy of the observations was compared with that of the Paris and San Fernando instruments. No substantial difference was found.

The mean residuals in zenith distance for stars in the program were analyzed for possible dependence on magnitude, spectral type, and azimuth. To investigate the effect of magnitude on mean residual the stars were divided into groups by half-magnitude. The mean of the mean residuals and its standard deviation were formed for each half-magnitude group. The standard deviation appears to be larger for brighter stars ($1.0 < m_v < 1.5$) and somewhat larger for faint stars ($6.0 < m_v < 6.5$). It is essentially constant for intermediate magnitudes. No correlations were found in a similar analysis for dependence of residuals on spectral type. A systematic trend appears in a correlation of residuals with azimuth. Those stars observed in azimuths between 0° and 90° and between 180° and 270° tend to be negative while those observed in the other quadrants tend to be positive.

After smoothing the observations in time and latitude using Vondrak's method, the systematic differences of each observer's results from the adopted curves were found. These personal equations may be as large as $0\mathrm{\overset{s}{.}}007$ in time and $0\mathrm{\overset{''}{.}}04$ in latitude. Systematic differences in time and latitude for each group in the observing program were also calculated in this manner. These group corrections were also determined using the chain method (Guinot, 1958) to give two separate estimates for group corrections. In time the corrections may be as large as $0\mathrm{\overset{s}{.}}022$ and the two estimates may differ by as much as $0\mathrm{\overset{s}{.}}020$. In latitude the group corrections may be as large as $0\mathrm{\overset{''}{.}}17$ and the two estimates may be different by as much as $0\mathrm{\overset{''}{.}}13$.

Values of UT2 - TAI derived from the observations at Merate were compared with UT2 - TAI of the Bureau International de l'Heure (BIH). The derived values were computed using the polar coordinates given in BIH Circular D and were smoothed by Vondrak's method. A preliminary evaluation shows the existence of an annual component in the Merate observations from 1971 through 1973 which diminishes in 1974 and almost disappears in the following years.

REFERENCES

Buffoni, L., Carta, F., Manara, A., Mazzoleni, F.: 1975a, Rapporti Interni Dell' Oss. Astr. di Brera-Milano no. 10 - 12.
Buffoni, L., Carta, F., Chlistovsky, F., Manara, A., Mazzoleni, F.: 1975b, Bolletino di Geodesia e Scienze Affini 34, no. 3, p. 249.
Guinot, B.: 1958, Bull. Astronom. 22, p. 1.

THE LONGITUDE DIFFERENCE MERATE - MILANO DERIVED FROM DANJON ASTROLABE
OBSERVATIONS BY MEANS OF AN ONE STEP ADJUSTMENT USING AN EXTENDED MODEL

Klaus Kaniuth
Deutsches Geodätisches Forschungsinstitut, München

Werner Wende
Bayerische Kommission für die Intern. Erdmessung, München

ABSTRACT

As a part of the establishment of a unified longitude system for the European Triangulation Network the difference in longitude between the reference points Merate and Milano was measured with a Danjon Astrolabe. This paper describes the results of a one-step adjustment of these observations including additional parameters for effects like personal equations and catalogue errors.

1. INTRODUCTION

A substantial basis for the new adjustment of the European Triangulation Network is a homogeneous net of astronomical longitude reference points. As a part of the establishment of such a reference system the differences in longitude between München (Germany) = geodetic reference point, Merate (Italy) = BIH station (MIA) and Milano (Italy) = geodetic reference point and former BIH station (MII) have been determined from measurements with a Danjon Astrolabe in München, Merate, Milano and again München at the end of 1977.

This paper describes the derivation of the longitude difference between Merate (MIA) and Milano (MII) from a common adjustment of all star transits observed at these two observatories in one step, in contrast to the usual way of groupwise adjustment and subsequent computation of the stations longitudes in a second step. The measurements were done by two observers who alternated with each other and who observed within 13 nights 65 groups with altogether 1271 stars.

2. ADJUSTMENT MODEL

The fundamental relation for the evaluation of equal altitude observations is the well known cosine theorem in the astronomic triangle. In a common adjustment of all observations at both stations it has to be solved for the following parameters:

Latitude $\varphi_i = \varphi_{io} + d\varphi_i$, longitude $\lambda_i = \lambda_{io} + d\lambda$,

difference of personal equations $d\varphi_p$ respectively $d\lambda_p$,
zenith distance $z_k = z_{ko} + dz_k$
with $i\,[1:2]$ = station index, $k\,[1:65]$ = group index.

In addition to these 71 fixed parameters we introduced into the model right ascension corrections $d\alpha$ which were set up iteratively for those stars whose mean residuals applying the t-test turned out to be significant on the 95% level. Thus the observation equation for a star transit s becomes

$$v_s = -\cos a_s\, d\varphi_i - \frac{j-1}{2}\cos a_s\, d\varphi_p - \cos\varphi_{io}\sin a_s\, d\lambda_i \\ - \frac{j-1}{2}\cos\varphi_{io}\sin a_s\, d\lambda_p - dz_k + (\cos\varphi_{io}\sin a_s\, d\alpha_r) - l_s \quad (1)$$

with

a_s = north azimuth, $j[1,3]$ = observer index,
r = index of stars for which corrections $d\alpha$ had to be estimated,
l_s = free term computed in the usual way from the observed time of transit, approximate values of the unknowns and the apparent places; l_s includes a refraction term and is corrected for polar motion using the pole coordinates and UT1-UTC published by the BIH in circular D.

Introducing the notations

$$x_f^T = (d\varphi_1, d\varphi_2, d\varphi_p, d\lambda_1, d\lambda_2, d\lambda_p, dz_1, \ldots, dz_{65}),$$
$$x_v^T = (\ldots, d\alpha_r, \ldots),\quad v^T = (v_1, \ldots, v_{1271}),\quad 1^T = (1_1, \ldots, 1_{1271}), \quad (2)$$

B_f = coefficient matrix of the fixed parameters x_f,
B_v = coefficient matrix of the variable parameters x_v,
Q_1 = variance-covariance matrix of the observations,

the system of observation equations has the least squares solution

$$\begin{pmatrix} x_f \\ x_v \end{pmatrix} = [(B_f, B_v)^T Q_1^{-1}(B_f, B_v)]^{-1}(B_f, B_v)^T Q_1^{-1} 1. \quad (3)$$

The normal equation system is partitioned into a fixed and a variable part in order to have not to invert the whole system again in each iteration. The variable part is successively extended until no further star turns out to be in need of a right ascension correction $d\alpha$. With the abbreviations

$$N_{ff} = B_f^T Q_1^{-1} B_f, \quad N_{fv} = B_f^T Q_1^{-1} B_v, \quad N_{vv} = B_v^T Q_1^{-1} B_v,$$
$$Q_{ff,1} = N_{ff}^{-1}, \quad Q_{vv} = (N_{vv} - N_{fv}^T Q_{ff,1} N_{fv})^{-1}, \quad (4)$$
$$Q_{fv} = -Q_{ff,1} N_{fv} Q_{vv}, \quad Q_{ff,2} = -Q_{ff,1} N_{fv} Q_{fv}^T,$$

one gets

$$x_f = Q_{ff,1}^T \; B_f^T \; Q_1^{-1} \; 1 + \underbrace{Q_{ff,2}^T \; B_f^T \; Q_1^{-1} \; 1 + Q_{fv}^T \; B_v^T \; Q_1^{-1} \; 1}_{\text{variable}} \tag{5}$$
$$ \underbrace{\phantom{Q_{ff,1}^T \; B_f^T \; Q_1^{-1} \; 1}}_{\text{fixed}}$$

$$x_v = Q_{fv}^T \; B_f^T \; Q_1^{-1} \; 1 + Q_{vv}^T \; B_v^T \; Q_1^{-1} \; 1.$$

The variance-covariance matrix Q_1 of the observations is not known a priori but one can assume that the covariances are negligible and that the variances within one group are equal because they seem first of all to depend on the weather conditions and on the observer's disposition. Under this prerequisite Q_1 is diagonal and its elements q_k can be estimated from the observations in an iterative process (Kubik 1967). Starting from any, for instance equal variances for all groups one gets new estimates q_k of the variances after each adjustment from the equations.

$$q_k = \frac{1}{n_k} v_k^T v_k - b_k, \quad b_k = -\frac{1}{n_k} \text{tr}[(B_f, B_v)_k Q_x (B_f, B_v)_k^T] \tag{6}$$

with $\quad n_k$ = number of stars observed in group k,

v_k = subvector of v

$(B_f, B_v)_k$ = submatrix of (B_f, B_v) $\Big\}$ belonging to group k.

3. RESULTS

According to the given formulae system a common adjustment of all 1271 observed star transits was made, and that in two versions:

A. An adjustment without iterative estimation of the 65 group variances. In this case of assuming equal weights for all observations the computer program made 42 iterations for deriving corrections $d\alpha$ of 22 FK4 and 19 FK4 Sup stars.
B. An adjustment including the estimation of the group variances from the observations themselves according to equations (6). In this case the program needed two iteration steps for the estimation of variances which changed no more significantly in a further iteration. Within one variance iteration 47 iterations for deriving corrections $d\alpha$ of 25 FK4 and 21 FK4 Sup Stars have been computed.

The different number of iterations in the two adjustments is due to the different variance matrices of the observations. As main results of the two adjustments the longitudes of Merate (MIA) λ_1 and of Milano (MII) λ_2, the difference of personal equations $d\lambda_p$ and the derived longitude difference $\Delta\lambda$ with their r.m.s. errors are given in table 1.

From the observations at the reference point München done before and after those at the Italian stations it was evident that the personal equations in longitude did not alter.

Parameter	A	B
λ_1, m_{λ_1}	$37^m 42^s\!.7856 \pm 0^s\!.0012$	$37^m 42^s\!.7833 \pm 0^s\!.0011$
λ_2, m_{λ_2}	$36\ 45.8322 \pm 0.0012$	$36\ 45.8314 \pm 0.0009$
$d\lambda_p$, $m_{d\lambda_p}$	$-\ 0.0069 \pm 0.0014$	$-\ 0.0067 \pm 0.0012$
$\Delta\lambda$, $m_{\Delta\lambda}$	$56^s\!.9534 \pm 0^s\!.0013$	$56^s\!.9519 \pm 0^s\!.0011$

Table 1: Results of the Adjustment

The conventional longitude used in the BIH 1968 reference system was derived for Milano (MII) from observations in the period 1966.50-1967.45, whereas that of Merate (MIA) was established by the observatory and was not based on astronomical observations (Guinot 1978). We have assumed that the actual longitudes of these stations may be obtained by adding the term a' determined by the BIH for the last year of operation; the value is given for MII in the BIH Annual Report for 1969, and for MIA in 1977 by Guinot (1978). The resulting longitudes are:

Merate (MIA) $\lambda_1 = 37^m 42^s\!.7665$ Milano (MII) $\lambda_2 = 36^m 45^s\!.8359$

Thus the difference in longitude in the BIH system between these two stations is approximately $56^s\!.931$.

Summarizing the results of this paper and comparing them with those given in the BIH system one may make the following statements:

- The common adjustment of all observations using the described model proved useful. The estimation of the group variances within the adjustment yields a remarkable improvement in the inner accuracy of the derived parameters.
- The longitudes of Milano (MII) agree fairly well within 4 ms. In the case of Merate (MIA) there results a difference of about 17 ms which is however imaginable if one considers that the r.m.s. error of coefficient a'$_{1977}$ is $\pm$ 7.9 ms (Guinot 1978).
- It should be considered whether longitude differences like the observed one with accuracies of a few ms could be used for an improvement of the BIH longitude reference system.

4. REFERENCES

Guinot, B.: 1978, private communication.

Kubik, K.: 1967, "Zeitschrift für Vermessungswesen" 92, pp. 173-178.

DISCUSSION

S. Debarbat: Am I correct in my deduction, from the small errors that you quote, that the Danjon astrolabe is a very suitable instrument for the determination of differences of longitude?

K. Kaniuth: Yes, in my opinion it is. But I should add that in the case of Merate and Milano, because of the small latitude difference between the stations, nearly identical star programs could have been observed.

VELOCITY OF THE MOTION OF THE TERRESTRIAL POLE

Janusz Moczko
Astronomical Latitude Observatory
Polish Academy of Sciences
Borowiec/Kornik, Poland

ABSTRACT

The results of the spectral analysis of the velocity of the polar motion based on the data of the International Latitude Service (ILS), International Polar Motion Service (IPMS), Bureau International de l'Heure (BIH), and the Doppler Polar Motion Service (DPMS) are presented.

1. INTRODUCTION

Since the polar coordinates are computed for equally spaced time intervals, n-1, n, n+1, etc., the velocity may be represented by the distance, S_n, the pole moves during these intervals. Thus the minimum velocity during the time interval, n, denoted by V_n can be given by

$$V_n = S_n = [(x_n - x_{n-1})^2 + (y_n - y_{n-1})^2]^{\frac{1}{2}}.$$

2. OBSERVED VELOCITY

Four different sets of polar coordinates were analyzed:

    a. BIH five-day values from 1967 through 1976 (Annual Report of the BIH),
    b. ILS, IPMS, BIH values for every twentieth of a year from 1962 through 1967 (Annual Report of the BIH, Annual Report of the IPMS),
    c. DPMS five-day values from 1972 through 1976 (U. S. Naval Observatory Time Service Pub. Series 7),
    d. ILS and BIH values for every tenth of a year from 1900 through 1976 (Annuaire 1974 du Bureau des Longitudes, Annual Report of BIH).

The velocities computed for the five-day intervals are very noisy. The velocities derived from the other data (Figures 1 and 2) show that there are no Chandler and annual periodic variations in the velocity but there is a variation with a period of six or seven years. The variation in velocity is quite similar for the ILS, IPMS, and BIH data.

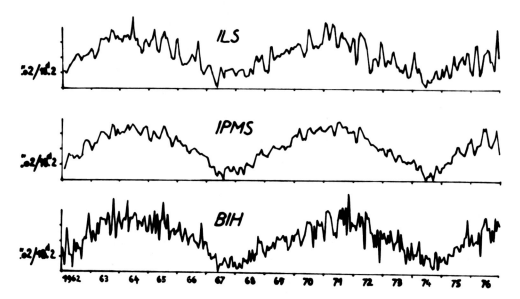

Figure 1. Rate of polar motion in twentieth of a year intervals 1962-1976.

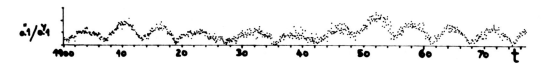

Figure 2. Rate of polar motion in tenth of a year intervals 1900-1976.

3. SPECTRAL ANALYSIS

The solid-line spectra in Figures 3 and 4 show the results of the spectral analyses of the V_n data. To avoid the influence of the low-frequency terms in the spectra it was decided to perform a similar analysis on the first differences of the V_n data, $\Delta V_n = V_n - V_{n-1}$. Physically these data are proportional to the accelerations of the polar motion. Periodic variations in V_n should also be present in the ΔV_n data, but the low frequencies will be damped.

The results of the spectral analyses of the ΔV_n are represented in Figures 3 and 4 by broken lines. The similarity in the periodicities found in the spectra of the V_n and ΔV_n is apparent. The enhancement of the higher frequencies in the spectra of the ΔV_n data is also noticeable.

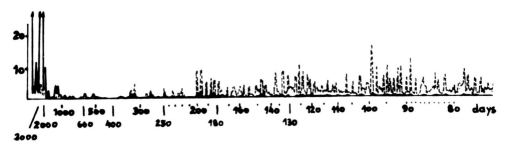

Figure 3. Spectral analysis:
  a. (solid line) rate of polar motion in tenth of a year intervals 1900-1976;
  b. (broken line) first differences of the rate of polar motion in tenth of a year intervals 1900-1976.

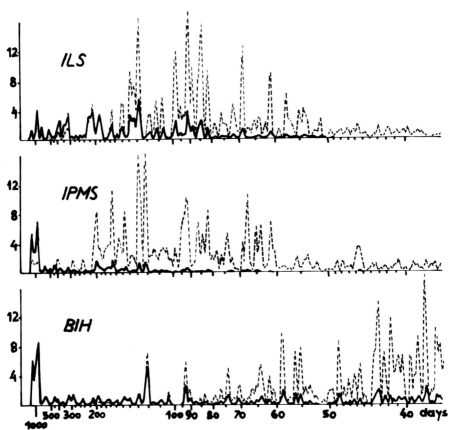

Figure 4. Spectral analysis:
  a. (solid line) rate of polar motion in twentieth of a year intervals 1962-1976;
  b. (broken line) first differences of the rate of polar motion in twentieth of a year intervals 1962-1976.

4. CONCLUSIONS

This spectral analysis of the velocity and acceleration of the polar motion allows us to draw the following conclusions.

1. The main components of the polar motion (Chandler, annual, and semi-annual) have constant velocities.
2. Distinct variations of the velocity with periods of 6.0 and 6.8 years are present. These may be due to superpositions of annual and Chandler periods as well as semi-monthly and semi-annual nutation components.
3. High-frequency components with periods of 320, 190, 220, 120, 90, 27, 18, 14, and 11 days are less distinct. These may be explained by the superposition of nutation components (Table 1).
4. Strong disturbances of the Earth's rotation occurred from 1918 to 1930. Separate spectral analyses of V_n for the periods 1900-1919, 1917-1936, and 1932-1977 show that during the period 1917-1936 the six-year components disappeared and a ten-year periodicity was present.
5. It is advantageous to analyze the velocity of the polar motion because of its independence from the reference system.

Table 1. Periods of superimposed nutation components.

Period (days)	Contributing Nutation Term Periods (days)
1030	182.62, 5.64
327	13.66, 26.94
202	13.66, 14.76
186	13.66, 13.63
131	13.66, 9.56
125	13.66, 9.13
97	13.66, 7.10
94	13.66, 6.86
77	13.66, 5.64

REFERENCES

Annuaire 1974 du Bureau des Longitudes: Paris.
Annual Report of the Bureau International de l'Heure for the years 1967-1976: Paris.
Annual Report of the International Polar Motion Service for the years 1962-1974: Mizusawa.
U. S. Naval Observatory Time Service Publication Series 7, 1972-1976: Washington.

DISCUSSION

D. Djurovic: Have you identified a term which has a period close to 122 days in the latitude spectrum?
B. Kolaczek: Yes.

PART III : REFERENCE SYSTEMS

THE REFERENCE SYSTEMS

by J. Kovalevsky
C.E.R.G.A. Grasse, France.

ABSTRACT. In order to discuss accurately the motions of the Earth in space, it is necessary to define rigorously two readily accessible reference systems. The conception and the realization of celestial absolute systems and terrestrial coordinate systems are discussed. It is suggested that these systems of reference ought to be defined with a minimum of theoretical or observational constraints. Examples of such ideal reference systems are given, together with some desirable properties for intermediate systems.

INTRODUCTION

"Toute l'astronomie repose sur l'invariabilité de l'axe de rotation de la Terre à la surface du sphéroide terrestre et sur l'uniformité de cette rotation".

If this statement by P.S. de Laplace (1825) still held, there would be no IAU symposium on "Time and the Earth's rotation". But since we know that it is not true, there arises the problem: since the description of the motion of a body is possible only with respect to something else, what shall we refer the motion of the Earth to?

As a matter of fact, we need two systems of reference: a terrestrial coordinate system that would represent the body "Earth" and to which observatories as well as the axis of rotation are referred, and an absolute external celestial system of reference in which the motion of the first system represents what we call the "Rotation of the Earth" - that is, the motion of the Earth around its centre of mass.

Many different definitions of the two systems are possible and some are reviewed in this presentation. They are not equivalent and may have different characters as far as their conceptual simplicity, their practical realization or their accessibility are concerned. Some may be practical for some kind of observations and completely unfit for others. Let us discuss these different points for the celestial system.

1. CELESTIAL REFERENCE SYSTEMS

1.1. Dynamical systems

The main requirement for a celestial reference frame is that it should be inertial. By this, we mean that there exists no residual rotation of the system. Strictly speaking this requirement is a dynamical one and it implicitly assumes the validity of Newtonian mechanics, corrected if necessary for relativistic effects. If we call (S_1) such an ideal absolute frame of reference, and if we assume that $\vec{\omega}$ is the rotation vector of another system (S) with respect to (S_1), then the absolute acceleration of a point P, $\vec{\Gamma}_1$, differs from its acceleration in the system (S) by:

$$\delta\vec{\Gamma} = \vec{\Gamma}' + 2\vec{\omega} \wedge \vec{V}' \tag{1}$$

where $\vec{\Gamma}'$ and $\vec{V}'$ are the acceleration and the velocity of (S) relative to (S_1).

The quantity $\delta\vec{\Gamma}$ enters in the differential equations of the motion of P as referred to (S). The condition that (S) is an absolute system is:

$$\delta\vec{\Gamma} = 0 \tag{2}$$

So the detailed analysis of the motion of a system of celestial bodies, like the Moon or the components of the solar system, may provide corrections that would make the chosen reference system absolute, by determining the parameters of equation (1). Hence, this provides an access to the absolute reference system.

The complete solution of the equations of motion may, however, contain terms having a structure similar to the solution of equation (1). For instance, if a single planet is taken as the material system, the mean motion n of the planet around the sum is linked to the semi-major axis a by Kepler's third law:

$$n^2 a^3 = km \tag{3}$$

If there are errors Δa and Δkm in the assumed values of a and km, one has the following relation:

$$\frac{2 \Delta n}{n} + \frac{3 \Delta a}{a} - \frac{\Delta km}{km} = 0 \tag{4}$$

and this implies that Δn and ω cannot be determined separately; the absolute system therefore cannot be derived from the measurement of n. In the strict case of the Newtonian two-body problem, the absolute system has to be defined by the condition that the pericenter is fixed in space. But in relativistic celestial mechanics this is no more true, and one must also know the exact value of the parameter of the Schwartzschild model in order to have access to the absolute frame. For the Moon, a similar difficulty arises since, in addition to the Newtonian accelerations due to the Earth, the Sun and the planets, and to the relativistic effects, the existence of a poorly modelled secular acceleration introduces new difficulties in determining $\vec{\omega}$.

In practice, more complex dynamical systems, including many bodies like the massive components of the solar system, are used. The complexity of the system appreciably decorrelates the equations in $\vec{\omega}$ from other unknowns, but there are proportionally many more parameters to be determined from the observations in order to solve the whole system of equations. So, finally, the actual determination of the reference frame is not necessarily improved.

These examples show that the definition of a system of reference implies the existence of a model of the physical system that is used to define it. I have described elsewhere some of the models associated with dynamical systems (Kovalevsky, 1975). Other examples of the complexity of the parametrization of the material system may be found in Mulholland (1977) for the lunar motion, in Duncombe et al. (1975) in the case of the solar system and in Anderle and Tanenbaum (1975) in the case of a system defined by the motion of artificial satellites.

These examples show clearly that a dynamical system of reference having accuracy in the range of 0''.1 to 0''.01 can only be defined through a complex model incorporating many parameters which must be determined simultaneously from the observations. For example, almost all the system of astronomical constants (see Müller and Jappel, 1977), and a set of six mean or instantaneous orbital elements for each planet and for the Moon, are necessary to define and give access to a reference system based on the planetary system; the complexity would need to be even greater to yield the desirable accuracy of 0''.001.

1.2. Kinematic systems

Another approach to the definition and the realization of a celestial reference frame is to consider that point-marks of the reference systems are distant celestial bodies, stars or galaxies. In the Universe, light follows well defined geodetic lines, the apparent directions of which can be easily reduced to the actual directions by appropriate aberration corrections; but since celestial bodies are not fixed, one also needs some model of their motion with respect to what we believe is an absolute reference system. A model of the distribution of proper motions has hence to be adopted. And here lies a major difficulty. If we consider the case of stars, there is no means of completely separating a general rotation of the system of stars (galactic rotation) from the rotation of the reference frame. Therefore, in the construction of a representation of the reference system by a catalogue of star positions and proper motions (like the FK 4 or the future FK 5), it is necessary to use a dynamical definition of the reference frame. Classically, this is done by the introduction of the equator and the ecliptic as reference planes obtained through the observations of the position of the Sun and the planets, and a model of their respective motions through theories of the rotational behaviour of the Earth (precession, nutation), and the motion of the Earth and planets around the Sun (planetary precession, motion of the ecliptic). This leads to the very difficult and delicate task of fitting the star system to the assumed motions of these planets (Fricke, 1974 and 1975).

So, at present, there does not exist any purely geometric reference system. However, if one assumes that extragalactic point sources have very small relative proper motions and that there is no transverse component in the expansion of the Universe, then we obtain a static reference system. This is a particularly simple model and, conceptually, an ideal one.

1.3. Practical realizations

Among these various possible types of reference system, which are those that are the most advantageous for the study of the Earth rotation? The reply to this question depends upon the intrinsic properties of the system, but also upon its accessibility to the type of instrument that is to be used to measure the Earth rotation parameters.

1.3.1. Laser techniques on the Moon

Laser ranging, to the Moon or to satellites, provides a very promising technique for determining UT1 and polar motion. These are obtained by analysis of the residuals obtained at several stations when nominal rotation parameters are used in the process of global determination of all the quantities involved in the physical description of the motion of the satellite (see, for instance, Kolenkiewicz et al., 1977) or the Moon (Harris and Williams, 1977). The determination of the absolute reference system is a part of the global dynamical discussion of the motion of the body and is one of the limiting factors in the precision of the results.

In the case of the Moon, the main references are the ephemeris of the centre of mass of the Moon and the ephemeris of the rotation of the Moon about its centre of mass. The construction of both ephemerides is subject to errors of modelling and, in particular, to errors in the values of the higher moments of the lunar gravitation field, the knowledge of the free libration of the Moon (Calame, 1976) and the inadequate modelling of the transfer of angular momentum due to the tidal dissipation of the Earth. Most of these parameters are gradually improved as more and more lunar ranging observations are included in the solution defining the ephemeris. Gradually, then, these ephemerides define a better absolute reference system, though it

THE REFERENCE SYSTEMS

is difficult to say how much the outcome is free from a residual rotation and possible terms of long period. The short periodic terms are certainly well represented, so this system is fit for short range Earth rotation determinations. The worst part of the model is probably due to the dissipation terms. However, even those can be determined from lunar ranging with reasonable accuracy, and residual accelerations of the system are of the order of magnitude of the error in the determination of the secular acceleration of the Moon, at present about 5" per century square.

This system is a self-defined system that can be used only for lunar laser ranging. It cannot be used by other techniques unless other observations link it to another reference system. This is done for a star defined system like the FK4 using meridian observations and occultations by the Moon. But this procedure degrades the initial precision of the dynamical system, since the supplementary observations are not as precise as lunar ranging and, at the best, the residual long term drift of the transferred system will not be better than that of the comparison.

In conclusion, this system is certainly very good for short and medium range in time (say 10 years). The residual rotation (or accelerations) seems to be a limitation that can be reduced only by comparison with a better known reference frame.

1.3.2. Laser techniques on artificial satellites

In the case of satellites analogous difficulties arise, but in a much shorter timescale. For instance, the longitude origin is essentially arbitrary and many forces which cannot be accurately modelled act on satellites to give effects that exceed the acceptable limits in a few weeks or, at the best, a few months (Anderle et al., 1975). Imperfections in the models used for calculation of solar and terrestrial radiation pressure and for the representation of the Earth's gravity field produce estimated uncertainties of a few centimeters per week even for LAGEOS, the satellite that has been most nearly optimised in these respects. The laser observations therefore do not provide full orientation in inertial space and the accumulated effects may reach 0".01 per year, which is worse than the present knowledge of the precession.

Another drawback of this system is that it is specific to the laser observations of a given satellite and, unlike that based on observations of the Moon, it cannot be tied to other systems. So a dynamical system defined by artificial satellites is accurate only over a short period of time. It can be useful for the measurement of transient and short-period phenomena, but it is not fit for middle-term and long-term studies.

1.3.3. Stellar systems

These are well known and will not be analysed here (see Fricke 1974). The best examples are the FK 4 and the FK 5 systems, the latter being now under construction (Fricke 1975). As we have said, they are essentially dynamical and are based on dynamical models of the Earth and of the solar system. The dynamical discussion of the residual rotation in longitude (Laubscher, 1976) showed a residual rotation in the system that is determined to 0".1 per century. There is also a residual drift of 0".3 per century in the obliquity that exists in all planetary observations. It has been suggested (Duncombe and Van Flandern, 1976) that this could be an artificial effect of an incomplete reduction to a fundamental star reference system. This is not the place to discuss this assumption, but whatever is the reason, it is a good illustration of the limitations in the construction of a classical system of reference.

One can describe a stellar reference frame by the following characteristics:

- It is easily accessible through observations of bright stars (m < 10), planets or the Moon;

- Its long term precision is limited to about ± 0".1 per century;

- It has regional errors (± 0".05 in position and ± 0".003 per year in proper motions for the FK 4 and, it is hoped, about half these values in the FK 5). These errors may introduce biases that are very difficult to analyse in the motions referred to the system; the problems are well known to all users of such catalogues;

- Its extension to higher magnitudes (12 or more) is possible only through photographic observations that significantly alter the accuracy of the realization.

1.3.4. Extra-galactic radio system

Such a system would be defined by the fixed positions of a few extra-galactic radio-sources whose positions are determined by very long base interferometry. The relative accuracy of these positions represents the accuracy of the system. And since they are fixed sources, there is no rotation or drift of the system. Actually this relies on the assumptions that the sources have no transverse velocities and are physically well defined, and are not, for instance, transient patches of synchrotron radiation. This requirement reduces the already small number of accessible sources. However, it is not too optimistic to consider that there might be over 20 reasonably bright sources having these qualities that could be the basis of a

THE REFERENCE SYSTEMS

reference system. A list of such sources was proposed by Elsmore and Ryle (1976). The choice of the spherical coordinate system in which these coordinates are expressed is completely irrelevant. It could be related in some way to right ascension and declination, but could equally well be completely disconnected.

The adopted final coordinates of the sources may be obtained by some kind of weighted compensation of the angular distance between sources, as obtainable from VLBI observations. The final outcome is a base B formed of three orthogonal unit vectors ($\vec{e}_1$, $\vec{e}_2$, $\vec{e}_3$). The direction of each source S_i is given by a unit vector $\vec{S}_i$ expressed in the base B.

Any other object that can be observed with respect to at least two of the basic sources of S_i S_j, has a direction defined by a unit vector $\vec{V}$ such that ($\vec{V}$, $\vec{S}_i$) and ($\vec{V}$, $\vec{S}_j$) are equal to the observed values. So, the densification of the catalogue representing the system is possible using only relative observations.

The possibility of using such a system was demonstrated theoretically by Walter (1974) and practically by Elsmore (1976) who made first - epoch observations for the determination of the constant of precession. He used 5 km base radio-interferometric observations of ten extra-galactic radio-sources that played the role of the markers of the absolute system.

However, the densification of this catalogue will be very difficult, since there exist only very few radio stars and well defined points like extra-galactic radio-sources. But the situation might be completely different if an astrometric satellite is launched.

1.3.5. The astrometric satellite

Such a satellite has been imagined by Lacroute and presented several times (Bacchus and Lacroute, 1974; ESRO, 1975) and recently studied by ESA as a phase A project (ESA, 1978). It will permit, if it is successful, to measure with an accuracy of 0".002 the positions of 100 000 stars and, in a single mission, their proper motions to 0".002 per year. A second mission 10 years later would improve the proper motions by another factor of 5.

The star catalogue that will be obtained will be free of all systematic regional distortion, because stars separated by about $70°$ are directly connected one to another in such an intermingled manner, that no regional shift may occur. However, the whole system may have a residual rotation.

Three new techniques that are now being developed should provide links between objects of magnitude 16-19 and stars within about $0°.5$ of them to an accuracy of 0".001; these are optical interferometry, now being tested with small telescopes by Labeyrie in CERGA, Connes' techniques using photographic masks for small fields, and the large space telescope. So we may expect that, ten years from now, we shall be able to tie the basic extragalactic sources defining the VLBI absolute system to stars observed by space astrometry techniques. The stellar catalogue of the astrometric satellite will also include some radio stars observable by VLBI and it is in this form that the VLBI system may become fully accessible.

2. TERRESTRIAL REFERENCE SYSTEMS

The terrestrial reference systems play two roles:

- Their first objective is to represent unambigously the position of points on the Earth.

- Their second objective is to represent the Earth as a whole, so that their rotational motion with respect to the celestial system represents the angular motion of the Earth.

The problem is that these two objectives are not fully compatible and therefore different reference systems have been used in each case.

2.1. Geodetic data

The objective of geodetic data is to give a unique computational procedure for defining the coordinates of a point on the Earth. The definition is very complicated and involves the introduction of a reference ellipsoid defined by its size, flattening, and an initial point of triangulation. Provisions are made that this ellipsoid has axes directed to the CIO and the Greenwich mean astronomical meridian.

However, the necessary reduction to the Geoid, the complex problem of the deflection of the vertical that permits linkage between astronomical and geodetic coordinates, and numerous other corrections that must be applied to the observed quantities, require the use of models of the local variations in the density of the crust. These drawbacks are well known to geodesists and have been reviewed recently by Mueller (1975) who clearly shows that the existing geodetic data are not fit to serve as acceptable terrestrial reference frames.

However there exists a conceptually much simpler solution: this is the world-wide geometric datum. It would consist, if the Earth could be considered as a rigid body, of a constellation of first order geodetic points with given observed rectangular coordinates in a single system, preferably geocentric. This is possible already for a few stations in the world that could have their relative positions determined by several techniques like satellite laser, lunar laser and VLBI. The coordinates of these stations would be treated like those of the radio-sources in the celestial VLBI system. Such systems

THE REFERENCE SYSTEMS

already exist in principle as the world geodetic network or as parts of dynamical geodetic models (SAO Standard Earth, GEM, GRIM, etc...). They may be greatly improved by using lunar laser ranging and VLBI and having a geophysically tested model of tidal displacements.

As these stations are on the real non-rigid Earth subject to plate motions, one will have to associate a "proper motion" with each point and treat the whole system of points as a catalogue of stars with proper motions but with, also, an undeterminable general rotation of the whole system. As a matter of fact, only relative motions of the plates can be determined, and there is no geophysical phenomenon that could "nail" the system. It would probably be best to consider one plate as fixed and consider the motions of all the others with respect to it.

In conclusion, one would obtain, as on the celestial sphere, a base B' of three perpendicular unit vectors ($\vec{e}_1$, $\vec{e}_2$, $\vec{e}_3$'), and the position of a point B_i would be given as a unit vector $\vec{P}_i$ expressed in the base B', and a distance r_i to the origin.

For the determination of the coordinates of other points in this system, various space or Earth-based techniques can be used. It is also to be noted that, in this case, the directions of $\vec{e}_1$, $\vec{e}_2$, $\vec{e}_3$, are arbitrary and should not be defined otherwise than by the coordinates of the reference points. Any other definition would introduce other phenomena (as, for instance, the position at some instant of the axis of rotation) and would complicate the realization as well as the definition of the system.

2.2. Other terrestrial systems

Many systems that are used, or can be imagined, are not linked to practically accessible points, but to dynamically or kinematically defined axes. Examples are:

- The principal axis of inertia; the inertial tensor is diagonal in this system.

- The axis of figure, about which the moment of inertia is a maximum.

- The mean axis, defined in such a way that the total angular momentum of motion relative to the system is zero.

- The instantaneous rotation axis of the Earth.

None of these axes is both directly observable and fixed. So they have to be defined through a parametrized model of the gravitational field of the Earth, or need to be followed by continuous observations, and in the latter case one would again ask the question, with respect to what?

And, with the improvement of observations, these systems would tend to be modified with time. Furthermore, the reference to some initial position at time t_o of such or such axis would imply a continuous chain of measurements that will have to control the reduction of later observations to the reference system. This also is not advisable.

2.3. Ideal and practical systems

The most conceptually simple "ideal" reference systems that could be constructed in the foreseeable future are:

- A celestial reference system defined by VLBI and densified by the astrometric satellite.

- A terrestrial reference frame defined by a constellation of stations whose motions are referred to a given "origin" tectonic plate.

The rotation of the Earth would therefore be described by the rotation vector R(t,t') that superposes the base B'(t') on the base B'(t) as expressed in the system of reference defined by B.

$$B'(t) = R(t,t') \times B'(t')$$

The instantaneous rotation of the Earth at time t is the time derivative R'(t) of R(t,t') when t' → t (figure 1)

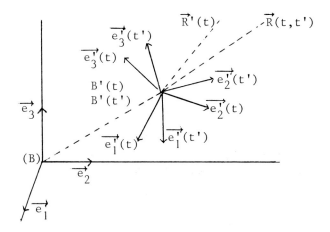

Figure 1. Bases and unit vectors of systems to be used for the rotation of the Earth

The direction of the rotation vector corresponding to R'(t) as measured in the base B' gives the motion of the pole and, in the base B, gives the direction of the instantaneous celestial pole.

The actual definitions of both bases B and B' may be completely arbitrary and probably ought to be so. For historical reasons they might be set close to the existing systems, but they certainly should not be set to coincide, not even for a given moment, since this would introduce new parameters to be more or less well determined and a model that would be more or less good. These factors would introduce an unnecessary inaccuracy in the realization of the systems. For the same reasons, no mention should be made of the equinox, which is defined by physical systems that are particularly complicated and difficult to model.

However, for practical reasons, several celestial systems may have to be used; for instance, a dynamical one for satellite work. It is possible to link these to the basic ones by measuring the "rotation of the Earth" by different methods referred to each of the systems. Let us call the bases of these systems B_1 and B_2. The rotation should be the same in both systems; if it is not then this defines the difference between B_1 and B_2 and, therefore, permits linkage of B_1 and B_2.

In some cases it may also be useful to introduce intermediate systems; for instance, a system in which the origin remains close to the equinox and the main circle is close to the equator of the Earth. Such an intermediate system should be simply and unambiguously related to the basic system by formulae that do not depend upon current observations or models. The transformation formulae should be given in full in a closed form as a conventional definition, and not as a consequence of a model that would need to be changed as knowledge of the underlying physics improved.

The same precautions should also be applied to any transformation of the basic geodetic system on the Earth. In particular, I would favour the total disappearance of the astronomical system of coordinates on the Earth and consider that each observatory is defined by its geodetic position in the terrestrial system and has a deflection of the vertical that is part of the instrument if the local vertical is used as instrumental reference to the observations. This would have the advantage of permitting periodic or secular variations.

BIBLIOGRAPHY

A number of papers quoted in this presentation were published in the Proceedings of the IAU Colloquium No. 26 on Reference Coordinate Systems for Earth Dynamics held in Toruń (26-31 August 1974), edited by B. Kolaczek and G. Weiffenbach and published by the Polish Academy of Sciences. This important book is referred below as "IAU coll. 26,

Toruń".

Anderle, R.J. and Tanenbaum, M.C.: 1975, IAU, Coll. 26, Toruń, p. 341.

Bacchus, P. and Lacroute, P.: 1974, in "New problems in astrometry", IAU symposium No. 61, Reidel Publ. Co., W. Gliese et al. editors, p. 277.

Calame, O.: 1976, The Moon, Vol. 15, p. 343.

Duncombe, R.L., Seidelman, P.K. and Van Flandern, T.C.: 1975, IAU Coll. 26, Toruń, p. 223.

Duncombe, R.L. and Van Flandern, T.C.: 1976, Astron. J., 81, p. 281.

Elsmore, B.: 1976, Monthly Notices Roy. Astron, Soc., 177, p. 291.

Elsmore, B. and Ryle, M.: 1976, Monthly Notices Roy. Astron. Soc, 174, p. 411.

ESA: 1978, "Space astrometry, Hipparcos", ESA document DP/PS(78), 13, 26 April 1978.

ESRO: 1975, "Space astrometry", Proc. of a symposium in Frascati, ESRO SP-108, March 1975.

Fricke, W.: 1974, in "New problems in Astrometry", IAU symposium No. 61, Reidel Publ. Co, W. Gliese et al. Editors, p. 23.

Fricke, W.: 1975, IAU Coll. 26, Toruń, p. 201.

Harris, A.W. and Williams J.G.: 1977, in "Scientific applications of lunar laser ranging", Reidel Publ. Co, Mulholland Editor, p. 179.

Kolenkiewicz, R., Smith, D.E., Rubincam, D.P., Dunn, P.J. and Torrence, M.H.: 1977, Phil. Trans. Roy. Soc. London, Vol. 284, p. 485.

Kovalevsky, J.: 1975, IAU Coll. 26, Toruń, p. 123.

Laplace, P.S. De: 1825, "Traité de Mécanique Céleste", tome 5, livre XI, p. 22.

Laubscher, E.: 1976, Astron. Astroph., 51, p. 9.

Mueller, I.I.: 1975, IAU Coll. 26, Toruń, p. 321.

Mulholland, J.D.: 1975, IAU Coll. 26, Toruń, p. 433.

Mulholland J.D.: 1977, in "Scientific applications of lunar laser ranging", Reidel Publ. Co, Mulholland Editor, p. 9.

Müller, E.A. and Jappel, A.: 1977, Proc. XVI-th General Assembly of the IAU, Vol. XVI-B, Reidel Publ. Co, p. 58.

Walter, H.G.: 1974, "Bulletin GRGS No 10", Meudon Observatory.

DISCUSSION

W.G. Melbourne: With regard to the inertial celestial system and to the use of very accurate optical astrometric positions of extragalactic radio sources, it should be noted that the centroids of these objects at microwave and visual wavelengths may differ by the order of 0.01 arsec, at least for those accessible with present VLBI technology.

P. Brosche: One should also use optical positions of galaxies, since quasars may have large apparent velocities and may show structure at the desired level of accuracy.

J. Kovalevsky: Yes, I agree. However, one should still have radio sources for the definition of the absolute system.

J.D. Mulholland: The estimate that photographic positions are limited to 0.1 arcsec is probably wrong, in that there will be a certain amount of astrometry on the Space Telescope.

THE EPHEMERIS REFERENCE FRAME FOR ASTROMETRY

C.A. MURRAY
Royal Greenwich Observatory

ABSTRACT. The basic problem of fundamental astrometry is to relate the instrumental reference frame of an observer to the frame defined by ephemerides of stars and objects in the Solar System. It is shown that in principle the choice of definition of the Ephemeris Reference Frame (ERF) can be quite arbitrary. For convenience, it is argued that the ERF should be defined by the rotation of an axi-symmetric rigid model Earth, the celestial pole being the direction of the axis of figure. This definition has practical and theoretical advantages over a model-dependent definition which attempts to take account of non-rigidity of the actual Earth.

The instantaneous ephemeris reference frame (ERF) is defined by the directions of the angular momentum vector of the Earth's orbital motion, and the celestial pole. This definition is embodied in the adopted numerical expressions for these two directions, relative to a supposedly inertial frame, and the ERF is rendered accessible to observation through ephemerides of stars and members of the solar system.

In order to compare observation with theory, the astrometrist needs to know, at any instant, the transformation between his own instrumental frame and the ERF. In the conventional language of meridian astronomy, he specifies the direction of the pole by the azimuth error of his instrumental collimation plane and the colatitude of his local vertical, and the direction of the equinox by his longitude or clock error relative to the ephemeris sidereal time.

The direction of the pole is determined by combining observations made at upper and lower culmination, either (a) of the same star or stars over an extended time interval, or (b) of different stars whose right ascensions differ by about 12 hours. A fuller discussion of the principles involved is given elsewhere (Murray 1978 b). These two procedures are superficially rather different. In case (a) the extended time interval must be at least 12 hours, for azimuth determination, and can extend to several years for colatitude. Any variation

of longitude, latitude and azimuth, due for example to thermal or
geophysical causes, should be known, and interpolated to each instant
of observation. In case (b) on the other hand, observations can be
made virtually simultaneously and only the differences of tabular
right ascension and declination, relative to ERF, have to be assumed.

Case (a) is best illustrated by considering colatitude. Variation
of latitude can be measured locally with a zenith instrument adjacent
to the meridian instrument, or obtained from the results published by
the BIH or IPMS. But observed variation of latitude itself depends on
differences of declination between stars, for example, in a zenith
zone, which have been determined by the chain or a similar method.
An exactly analogous procedure can be applied to variation of longitude,
but variation of azimuth is usually monitored by means of terrestrial
marks.

We therefore see that in principle, all fundamental meridian
observations are initially referred differentially to certain standard
stars, whose differences in right ascension and declination relative
to the ERF, are assumed known. Subsequently, the zero points in each
coordinate are determined from extended series of observations, including those of the Sun, Moon and planets.

Taking this procedure to its logical limit, we could take as
standards, stars in a zone very close to the pole, whose relative
positions could be mapped very accurately by photographic techniques,
thus avoiding problems of seasonal perturbations and closing errors
which are inherent in the chain method. The absolute scale of the map
must be derived from meridian observations, and is exactly analogous
to the determination of the zero point of the declination system. The
ephemerides of these stars, relative to the ERF, would give their
offset from the celestial pole, which would therefore be almost directly
observable at any instant.

It is important to note that at no point in this discussion has
it been necessary to appeal to any particular physical definition of
the celestial pole. Historically, this has been taken to be the direction of the angular velocity vector of a model Earth with rotational
symmetry, whose dynamical behaviour approximates closely to that of
the actual Earth. However, as Atkinson (1973, 1975) has pointed out,
this vector is essentially unobservable by the meridian astronomer,
and its use introduces small but troublesome short period variations
in the rotation of the ERF relative even to a rigid model Earth. Within
the framework of rigid dynamics, there is no doubt that the adoption
of the axis of figure as the definition of the celestial pole is
best for the observers.

However, now that the effects of departure from rigidity are
observable and can be modelled, there is a temptation to redefine the
ERF in terms of a more realistic model Earth. In the view of the
present author this would be a mistake. The dynamics of a rigid Earth

are well understood and the numerical representation of the rotation of such a rigid model is unlikely to require significant modification. On the other hand, an ERF which is severely model-dependent is liable to be changed as models improve. Deviations of the Earth from a rigid dynamical behaviour are, for the observer, compounded with purely local variations which must be observed directly anyway.

It is therefore proposed that the ERF should be defined by the instantaneous ecliptic and the axis of figure of an "Ephemeris Earth" which is a rotationally symmetrical rigid body with zero Eulerian motion; we may refer to the celestial pole of this reference frame as the "Ephemeris Pole". It has been shown elsewhere (Murray 1978 a) that the coefficients of the Oppolzer terms, representing the forced motion of the axis of figure relative to the angular momentum vector, which are given by Kinoshita (1977), should be used in preference to those listed by Woolard (1953).

The best available values of the displacement of the Ephemeris Pole from the direction of the axis of a more realistic model Earth should be made readily available; these should be regarded as estimates of corrections to be applied to observations in order to reduce them to the ERF. An observer would then have the option of either accepting these corrections or, alternatively, of deducing the direction of the Ephemeris Pole directly from his own observations. In this way the continuity of the ERF as the reference frame for astrometry can logically be preserved.

A further advantage of defining the ERF in this way, as has been pointed out elsewhere (Murray 1978 b), is that the component of angular velocity of the Ephemeris Earth about its axis of figure is rigorously constant, whereas the total angular velocity (about the axis of rotation) is not. We are thus led to a physically simple and rigorously self-consistent model for constructing not only the Ephemeris Reference Frame but also for defining a uniform standard for rotational time.

REFERENCES

Atkinson, R. d'E.: 1973, Astron. J. 78, pp. 147-151.
Atkinson, R. d'E.: 1975, Monthly Notices Roy. Astron. Soc. 171, pp. 381-386.
Kinoshita, H.: 1977, Celes. Mech. 15, pp. 277-326.
Murray, C.A.: 1978a, Monthly Notices Roy. Astron. Soc. 183, pp. 677-685.
Murray, C.A.: 1978b, Quart. J. Roy. Astron. Soc. 19, in press.
Woolard, E.W.: 1953, Astr. Pap. Wash. XV, part 1, p. 132.

DISCUSSION

F.P. Fedorov: I agree with Mr Murray in all points except one, which is one of terminology. It seems to me that the term "axis of figure" could cause confusion. That is why I used another term for the same axis; the "Jeffreys-Atkinson" axis.

C.A. Murray: I would not insist on the term "axis of figure", although it is exactly that axis of what I have called the Ephemeris Earth (to distinguish it from the real Earth) which is proposed as the definition of the celestial pole.

It should be pointed out that the pole which is currently used in the ephemerides is as logically distinct from the axis of rotation of the real Earth as the Jeffreys-Atkinson axis is from its axis of figure.

J.D. Mulholland: Is the avoidance of the rotational pole consistent with your concern for observational convenience?

C.A. Murray: Yes, certainly as far as classical methods are concerned. Indeed I believe that the only techniques for which the instantaneous axis of rotation has any relevance are those involving direct observation of velocity, such as Doppler measurements.

J. Kovalevsky: The reference frame proposed by Mr Murray is a fine example of what I called an intermediary system, since it is derived from the inertial frame by an unambigous mathematical formula independent of any possible modification introduced by a better understanding of the Earth's structure.

NUTATION IN SPACE AND DIURNAL NUTATION IN THE CASE OF AN ELASTIC EARTH

Nicole Capitaine
Observatoire de Paris

In order to improve the representation of nutation, the effect of elasticity of the Earth on the nutation in space and diurnal nutation of the terrestrial rotation axis is considered and its amplitude is evaluated for the principal terms. The choice between several methods taking this effect into account is discussed. A comparison with the effect induced on nutation by the existence of a liquid core in the Earth's interior shows that the consideration of elasticity alone cannot give any amelioration in the representation of nutation.

INTRODUCTION

For geophysical studies of the rotation of the Earth, it is necessary to take into account all effects acting on the motion of the instantaneous rotation axis which can be mathematically represented. The elasticity of the Earth is one of these effects.

The effect of elasticity on the free motion of the instantaneous rotation axis within the Earth is well known : it converts its Eulerian period of 305 days into the Chandlerian period of 430 days. The effect of elasticity on the tidal variation of gravity and on the tidal deviation of the vertical at each point of the Earth is also well known and is usually considered in the reduction of the observations.

But the modification of the coefficients of nutation due to elasticity is not well known and it is not clear if it is necessary or not to take it into account in the reduction of the observations. Poincaré (1910) and Jeffreys & Vicente (1957) have evaluated this modification as negligible. In contrast Fedorov (1963, 1977) and Mc Clure (1973) considered that it must be taken into account; however the modification proposed by these authors is not always correctly understood or not always admitted for the diurnal nutation (Mc Carthy 1976).

The question of the axes to which the coefficients of the nutation in space must be referred (Jeffreys 1959, Atkinson 1973) is also of great importance in the case of a non rigid Earth.

In this paper, we have computed the modifications of the coefficients of nutation, in the case of an elastic Earth, referred to three axes which can be considered : the Gz axis fixed within the Earth, the instantaneous axis of rotation and the axis of angular momentum. The results show how elasticity modifies the representation of nutation and whether or not it is necessary to take this modification into account.

1. SOLUTION FOR NUTATION IN SPACE AND DIURNAL NUTATION IN THE CASE OF AN ELASTIC EARTH

In the reduction of the observations, we have to express the position of a terrestrial reference frame (Gxyz) (G being the Earth's center of mass) with respect to a non-rotating reference frame. It is thus necessary to express the motion in space of the Gz axis. We do not consider here the Chandlerian free motion nor the forced annual motion of the rotation axis within the Earth, which are not predictable, so we limit the motion in space of the Gz axis to its lunisolar precession and nutation.

Considering the lunisolar torque and the variation of the Earth's inertia tensor due to rotational and tidal deformations, the complex solution for the coordinates of the pole of rotation in (Gxyz) can be written :

$$m = -iE_{\mathbb{C}} \sum_{j} R(j) A'_{21j} e^{-i(\omega_j t + \beta_j)} \quad (1)$$

with : $E_{\mathbb{C}} = \dfrac{3Gm_{\mathbb{C}}(C-A)}{c_{\mathbb{C}}^3 \Omega^2}$ (2), $R(j) = 1 - \dfrac{k_2 n_j}{k_s \Omega}$ (3), $A'_{21j} = A_{21j} / (1 - \dfrac{A}{C} \dfrac{n_j}{\Omega})$ (4)

G being the gravitational constant,
Ω the Earth's angular velocity, $n_j = \Omega - \omega_j$,
(A,A,C) the principal moments of inertia of the Earth in the non deformable case,
$m_{\mathbb{C}}$ and $c_{\mathbb{C}}$ respectively the lunar mass and the Earth-Moon mean distance,
k_2 and k_S respectively the Love number of degree 2 and the secular Love number,
A_{21j} the jth Doodson's real coefficient of degree 2 and order 1 for the total lunisolar term of frequency ω_j in the tidal potential,
$\omega_j t + \beta_j$ a linear combination of the mean sidereal time Φ, the mean lunar and solar longitudes ☉ and ☾, the mean longitudes of lunar and solar perigees p and π and the longitude ☊ of the lunar ascending node.

(1) is the expression of the diurnal nutation of the rotation axis within the Earth.

Let θ, Ψ, Φ be the Euler's angles between the (Gxyz) system and the non-rotating system (GXYZ) defined by the mean ecliptic and equinox of the epoch t_o : θ is the obliquity of the ecliptic, Ψ the longitude of the equinox and Φ the angle of rotation of the Earth. Using expression (1) for m and Euler's kinematical relations, we obtain, by an integration assuming $\sin \theta$ = constant, the following expression for the lunisolar precession and nutation of the Gz axis:

$$\Delta \theta_z + i\Delta\Psi_z \sin \theta = E_\mathfrak{C} \left[iA_{210}\Omega t + \sum_{j \neq 0} R(j) A'_{21j} \frac{\Omega}{n_j} e^{-i(-n_j t + \beta_j)} \right] \quad (5)$$

The motion of the Gz axis with respect to the (GXYZ) system can be obtained by using the expressions for the lunisolar precession and nutation of either the rotation axis or the angular momentum axis, respectively given by (6) and (8); the additional motions of these axes with respect to the Gz axis are given by (9) and (10) in the (GXYZ) system, and by (1) and (12) in the (Gxyz) system:

$$\Delta \theta_r + i\Delta\Psi_r \sin \theta = E_\mathfrak{C} \left[iA_{210}\Omega t + \sum_{j \neq 0} R(j) A''_{21j} \frac{\Omega}{n_j} e^{-i(-n_j t + \beta_j)} \right] \quad (6)$$

with: $\quad A''_{21j} = A_{21j}(1 - \frac{n_j}{\Omega}) / (1 - \frac{A}{C} \frac{n_j}{\Omega}) \quad (7)$

$$\Delta \theta_H + i\Delta\Psi_H \sin \theta = E_\mathfrak{C} \left[iA_{210}\Omega t + \sum_{j \neq 0} A_{21j} \frac{\Omega}{n_j} e^{-i(-n_j t + \beta_j)} \right] \quad (8)$$

$$\delta \theta_{rz} + i\delta\Psi_{rz} \sin \theta = E_\mathfrak{C} \sum_j R(j) A'_{21j} e^{-i(-n_j t + \beta_j)} \quad (9)$$

$$\delta \theta_{Hz} + i\delta\Psi_{Hz} \sin \theta = E_\mathfrak{C} (A/C) \sum_j R A'_{21j} e^{-i(-n_j t + \beta_j)} \quad (10)$$

with: $\quad R = 1 - \frac{k_2}{k_s} \quad (11)$

$$(H/C\Omega) = - iE_\mathfrak{C} R \sum_j A'_{21j} e^{-i(\omega_j t + \beta_j)} \quad (12)$$

2. COMPARISON WITH THE CASE OF A RIGID EARTH AND WITH THE CASE OF AN EARTH MODEL INCLUDING A LIQUID CORE

In the case of a rigid Earth, $k_2 = 0$, thus $R(j) = 1$ and $R = 1$.

In the case of an elastic Earth, as compared to the case of a rigid Earth, the amplitude of the circular nutation, j, in space of the Gz axis or of the rotation axis is multiplied by the factor $R(j)$. The corresponding modified coefficients of nutation in longitude and obliquity are computed in Table 1 for the principal terms. $R(j)$ is also the factor of modification, due to elasticity, of the amplitude of the term of frequency ω_j in the expression (1). This factor being between 0.94 and 1.06 for the considered diurnal waves, we see that the diurnal nutation of the rotation axis within the Earth is practically not affected by elasticity. This is confirmed by the values computed in Table 1.

In contrast, the motion in space of the angular momentum axis

Table 1 - Amplitudes (in 10^{-4}") of the coefficients ($\Delta\Psi\sin\theta, \Delta\theta$) of the principal nutations in space and of the corresponding terms of diurnal nutation within the Earth.

Argument	Axis of angular momentum			Rotation axis				Gz axis		
	Nutation in space	Diurnal Nutation		Nutation in space		Diurnal Nutation		Nutation in space		Jeffreys-Vicente's Earth-model
		Rigid Earth	Elastic Earth	Rigid Earth	Elastic Earth	Rigid Earth	Elastic Earth	Rigid Earth	Elastic Earth	
☊	-68722 92277	-2 12	-1 8	-68722 92277	-68726 92280	-2 12	-2 12	-68708 92267	-68713 92270	-68376 92020
2☊	827 -902	0 0	0 0	827 -902	827 -902	0 0	0 0	827 -902	827 -902	
☉ - π	501 0	1 1	0 0	501 0	501 0	1 1	1 1	501 -1	501 -1	
2☉	-5054 5509	-29 1	-20 1	-5054 5509	-5045 5501	-29 1	-29 1	-5084 5537	-5075 5528	-5249 5743
3☉ - π	-197 215	-2 0	-1 0	-197 215	-196 215	-2 0	-2 0	-199 217	-198 216	
g	269 0	5 5	4 3	269 0	269 3	5 5	5 5	269 -10	269 -7	
2α	-812 885	-67 2	-47 2	-812 885	-793 867	-67 2	-65 3	-881 949	-861 930	-903 978
2α - ☊	-136 183	-13 2	-9 1	-136 183	-132 180	-13 2	-12 2	-150 194	-146 191	
2α + g	-104 113	-13 0	-9 0	-104 113	-100 110	-13 0	-13 0	-118 126	-114 122	

remains unchanged in the case of an elastic Earth as compared to the case of a rigid Earth, but its diurnal nutation is multiplied by the factor R (R = 0.70 if k_2 = 0.30) . The corresponding values for the principal terms are given in Table 1 .

The modified coefficients of nutation of the Gz axis, computed by using the factors α(j) of Jeffreys&Vicente(1957) for the corresponding circular nutations in space of the rotation axis, in the case of an elastic Earth's model including a liquid core, are also given in Table 1 for the three principal terms. We can see that the modifications are much larger than in the simpler case of an elastic Earth.

The computations of Table 1 have been made by using the coefficients of nutation in space of the rotation axis given by Kinoshita (1977) for a rigid Earth and the values k_2 = 0.30, k_S = 0.96, (A/C) = 0.996.

CONCLUSION

Table 1 shows that :

- In the case of an elastic Earth, the rotation axis and the angular momentum axis are separated by 0".0020.

- The modifications of the coefficients of nutation due to elasticity are lower than 0".0001 except for the terms of arguments Ω, 2☾ -Ω, 2☾ + g for which it reachs 0".0004, for the term of argument 2 ☉ for which it reachs 0".0009 and for the term of argument 2 ☾ for which it reachs 0".0020 .

- If these modifications were the most important, the best method to take them into account must be to use the modified coefficients of nutation given in Table 1 for an elastic Earth and referred to the Gz axis , in order to avoid diurnal terms referred to the (Gxyz) system.

- Using the present IAU coefficients (Woolard 1953) referred (with a precision of 0".0001) to the axis of angular momentum and unchanged by elasticity, the simplest method to take elasticity into account has been given by Fedorov (1963, 1977) : it consists of multiplying the diurnal nutation of the axis of angular momentum for a rigid Earth by the factor R .

- The influence of elasticity on the coefficients of nutation is negligible with respect to the one due to the presence of a liquid core in the Earth's interior.

We can then conclude that the only good method to improve the representation of nutation must be to use the modified coefficients of nutation in space obtained from observations and from the most complete Earth's model.

The proposition of Atkinson (1973) consists to use the coefficients of nutation in space referred to the Gz axis. It seems that this method remains the simplest and the most satisfying one in the case of an elastic Earth and in the case of a more complete Earth's model.

REFERENCES

Atkinson,R.d'E.:1973,Astr.J.78,pp.147-151.

Fedorov,Ye.P.:1963,Nutation and Forced Motion of the Earth's Pole. Pergamon Press Ltd,Oxford,London,New-York,Paris.

Fedorov,Ye.P.:1977,to be published in the Proceedings of the IAU Symposium n° 78.

Jeffreys,H.,and Vicente,R.O.:1957,Mon.Not.R.astr.Soc.117,pp.142-161.

Jeffreys,H.:1959,Mon.Not.R.astr.Soc.119,pp.75-80.

Kinoshita,H.:1977,Celest.Mech.15,pp.277-362.

Mc Carthy,D.:1976,Astr.J.81,pp.482-484.

Mc Clure,P.:1973,GSFCG report X 592-73-259.

Poincaré,M.H.:1910,Bull.astr.Paris 27,pp.321-357.

Woolard,E.W.:1953,Astr.Pap.Amer.Eph.Naut.Almanac 15,1,pp.1-165.

CONCEPT FOR REFERENCE FRAMES IN GEODESY AND GEODYNAMICS: THE REFERENCE DIRECTIONS

Erik W. Grafarend*, Ivan I. Müller, Haim B. Papo, Burkhard Richter*
Dept. of Geodetic Science, The Ohio State Univ., Columbus, OH
*Inst. for Astronomical and Physical Geodesy, Univ. FAF at Munich, D-8014 Neubiberg

ABSTRACT

Modern high accuracy measurements of the non-rigid Earth are to be referred to four-dimensional, i.e., time and space-dependent reference frames. Geodynamics phenomena derived from these measurements are to be described in a terrestrial reference frame in which both space and time-like variations can be monitored. Existing conventional terrestrial reference frames (e.g., CIO, BIH) are no longer suitable for such purposes.

The ultimate goal of this study is the establishment of a reference frame, moving with the Earth in some average sense, in which the geometric and dynamic behavior of the Earth can be monitored, and whose motion with respect to inertial space can also be determined.

The study is conducted in two parts. In the first part problems related to reference directions are investigated, while the second part deals with positions, i.e., with reference origins. Only the first part is treated in this paper.

The approach is based on the fact that reference directions at an observation point on the Earth's surface are defined by four fundamental vectors (gravity, Earth rotation, etc.), both space and time variant. These reference directions are interrelated by angular parameters, also derived from the fundamental vectors. The interrelationships between these space and time-variant angular parameters are illustrated in a commutative diagram--tower of triads, which makes the derivation of the various relationships convenient.

In order to determine the above parameters from observations (e.g., laser ranging, VLBI) using least squares adjustment techniques, a model tower of triads is also presented to allow the formation of linear observation equations. Although the model tower is also space and time variant, these variations are described by adopted parameters representing our current knowledge of the Earth. For details, see <u>Bulletin Géodésique</u>, end of 1978.

PART IV : RADIO INTERFEROMETRY

AN INTRODUCTION TO RADIO INTERFEROMETRIC TECHNIQUES

B. Elsmore
Mullard Radio Astronomy Observatory
Cavendish Laboratory, Cambridge, England.

At this symposium we are to hear a great deal about new techniques for the measurement of earth rotation and polar motion that have come into being in the last decade and I am privileged to give a short introduction to one of these new techniques, that of radio interferometry.

The wavelengths currently used for high precision measurements of positions of sources in radio astronomy are typically 100 000 times longer than those of light. This immediately leads to the fact that in order to obtain the same theoretical resolution, limited only by diffraction, as that achieved by the classical optical instruments for measuring earth rotation, a radio telescope several kilometres in diameter would be required. Because of the impossibility of constructing such a large, single parabolic antenna, other means must be sought to obtain high resolution. This has led to the development of the radio interferometer which, in its most simple form, consists of two antennas spaced apart on the earth's surface. The two antennas are used simultaneously to observe the same source in the sky, the source position being derived from measurements of the difference between the radio frequency phase of the two signals received and from geometrical considerations as the earth rotates. The manner in which the two radio signals are compared has led to two distinct types of system. Details of both types are given elsewhere, see for example, the review of radio astrometry by Counselman (1976), and only some points will be mentioned here.

In the conventional interferometer system, known as the connected interferometer, the signals from the two antennas are brought together through cables or a radio link and are compared instantaneously. As the accuracy of the phase measurements depends upon the stability of the electrical length of the cables or link connecting the antennas to the phase comparison receiver, there is a practical limit to the maximum separation that can be successfully used; this amounts to about 10-20 kms using cables or a few hundred kms for a radio link.

To overcome this limitation, another interferometer type has been developed, known as VLBI, where the signals are tape recorded separately at the two sites, together with time marks, and at some later time the two recordings are brought together, are replayed and crosscorrelated. At first examination it seems that the use of larger and larger baselines operating at ever shorter wavelengths should provide more and more accuracy of measurements of source positions or earth rotation. An advantage of the VLBI system is that the baseline may be extended to several thousand kilometres, but there are some technical problems and the great potential of the VLBI system has yet to be fully realised in practice.

One difficulty is the very high stability demanded for the frequency standard at each observing station. The atomic hydrogen masers currently used have stabilities of 1 part in 10^{13} over a 10^3 or 10^4 second period, giving rise to phase instabilities at the signal frequency equivalent to a path length of 3 to 30 cms.

In the interpretation of all radio interferometric observations the observable quantity is the phase difference between the signals arriving at the two antennas, and hence there exists an ambiguity of $2\pi n$ of phase. Resolving this ambiguity is relatively simple in connected interferometers but is difficult in VLBI. It has been done in some VLBI applications by making, for example, simultaneous observations of an additional source with a pair of antennas at each end of a long baseline and using a common frequency standard to supply signals to both antennas at a given end. When observations cannot be phase connected unambiguously it is simple to measure the time derivative of the interferometer phase, giving rise to "fringe-rate" measurements. These have the disadvantage of being insensitive to declination near the equator and so to overcome this yet another technique has been developed (Rogers 1970) whereby wide radio frequency bandwidth observations are made and analysed to determine the group-delay which is effectively the derivative with respect to the angular frequency, $2\pi f$, of the phase difference.

Radio interferometers are affected by the atmosphere to a degree which depends partly on the baseline spacing. The steady atmosphere produces a delay which, for a source at the zenith, amounts to about 2 metres of extra path. It is sufficient with interferometers of small spacings to assume that the atmosphere above each of the antenna is substantially the same and hence only a small differential correction is needed to take account of the spherical structure of the atmosphere. This is not the case when spacings of thousands of kilometres are used and uncertainties of path length of the order 10-20 cms are then encountered. This uncertainty may be reduced by separate measurements made to determine the atmospheric path length at each site. In addition to the effects caused by the steady atmosphere, irregularities, mainly of water vapour in the troposphere, cause short term fluctuations of path length. One class of non-uniformity has a typical scale of

0.7 km and is associated with solar heating of the ground, whilst a second class has a scale 10-20 km and there is evidence that still larger scales exist (Hinder & Ryle 1971, Hargrave & Shaw 1978). It is these fluctuations that limit the positional accuracy attainable with the Cambridge 5 km connected interferometer. It is interesting to note that excellent conditions occur during widespread winter fog, when distortions to the incident wavefront may be less than 0.2 mm over 5 kilometres leading to a source positional accuracy of 0".02 arc; this illustrates a striking difference between requirements for optical and radio observations.

The effects of the ionosphere are in general less troublesome for two reasons; firstly, at short wavelengths the effects are small for moderate baselines and secondly, the delay introduced, unlike that due to the troposphere, is wavelength dependent and may therefore be eliminated by making observations at two wavelengths.

We are to hear later how determinations of positions of radio sources lead to measurements of earth rotation and polar motion and of the accuracies currently achieved. It should be noted that an interferometer orientated east-west cannot be used to measure polar motion but can provide measurements of UT1 without the knowledge of the instantaneous position of the pole (Elsmore 1973)

The radio sources themselves are fortunately distributed approximately uniformly across the sky, but they appear in all shapes and sizes with many of them having a double or complex structure. Only those sources that are compact, having an angular extent of less than about 0".2 arc, are ideal for astrometry or earth rotation measurements. These objects are nearly all extragalactic and include the very distant radio galaxies and quasars which provide an excellent inertial frame against which measurements of earth rotation may be made. So far radio astrometric measurements have been published for about 75 such sources in the northern hemisphere and whilst these must include the most intense sources, it seems probable that there may be a total of about 150 suitable sources in each hemisphere within reach of modest size antennas, i.e. one source every 140 square degrees. The highest precision claimed in recent surveys for measurements of right ascension are given in Table 1. It should be noted that the VLBI measurements are not significantly better than those of connected interferometers.

Table 1

Rogers et al. (1973)	VLBI	± 4 ms
Elsmore & Ryle (1976)	5 km interferometer	± 2 ms
Clark et al. (1976)	VLBI	± 2 ms
Wade & Johnston (1977)	35 km interferometer	± 2 ms

The radio emission from these objects is broad-band and, over the wavelength range 2-20 cms, currently used for high accuracy measurements, the emission is approximately uniform, apart from a tendency to be less at the shortest wavelengths. By way of a contrast, their optical emission is very feeble. Most of these objects have an optical magnitude in the range 16 to 20, which puts them out of reach of the conventional optical instruments used for the measurement of earth rotation.

In conclusion, it must be pointed out that radio interferometers provide an almost all weather, day and night facility for measuring earth rotation, but they are very expensive compared with, for example, a PZT, especially if we are considering the installation of an interferometer system at a site previously equipped only for optical observations. Of the two systems, the connected interferometer is cheaper and much simpler, but the VLBI system is capable of higher precision. It seems that interferometer measurements of earth rotation will only be made at existing radio observatories and speaking for one such observatory, it is disappointing that owing to the pressure of other commitments, it is not possible for us at present to make measurements on a regular basis. In addition to improved measurements with VLBI systems, I would also like to see a connected interferometer devoted entirely to astrometry and measurements of earth rotation.

References

Clark, T.A., Hutton, L.K., Marandino, G.E., Counselman, C.C., Robertson, D.S., Shapiro, I.I., Wittels, J.J., Hinteregger, H.F., Knight, C.A., Rogers, A.E.E., Whitney, A.R., Niell, A.E., Ronnang, B.O. and Rydbeck, O.E.H., (1976) Astron. J., 81, 599.

Counselman, C.C., (1976) Ann. Rev. Astron. Astrophys., 14, 197.

Elsmore, B., (1973) Nature, 244, 423.

Elsmore, B. and Ryle, M., (1976) Monthly Notices Roy. Astron. Soc., 174, 411.

Hargrave, P.J., and Shaw, L.J., (1978) Monthly Notices Roy. Astron. Soc. 182, 233.

Hinder, R. and Ryle, M., (1971) Monthly Notices Roy. Astron. Soc., 154, 229.

Rogers, A.E.E., (1970) Radio Science, 5, 1239.

Rogers, A.E.E., Counselman, C.C., Hinteregger, H.F., Knight, C.A., Robertson, D.S., Shapiro, I.I., Whitney, A.R. and Clark, T.A., (1973) Astrophys. J., 186, 801.

Wade, C.M. and Johnston, K.J., (1977) Astron. J., 82, 791.

DISCUSSION

A.R. Robbins: How do you establish the zero point of RA and connect it with that of the FK4?

B. Elsmore: Three of the surveys have used 3C273B to establish the zero point, adopting the position derived by Hazard et al. from lunar occultations. The Cambridge survey is based on the FK4 position of Algol; this incidentally gives a 6 ms difference when compared with the other surveys. The precision with which the zero point is established relative to an adopted source is better than 2 ms.

THE APPLICATION OF RADIO INTERFEROMETRIC TECHNIQUES TO THE
DETERMINATION OF EARTH ROTATION

Kenneth J. Johnston
E. O. Hulburt Center for Space Research
Naval Research Laboratory, Washington, D. C. 20375

Radio interferometric techniques for the determination of Earth rotation are described.

INTRODUCTION

The effect of variations in the Earth's rotation and polar motion in radio interferometry is to alter the orientation of the baseline, i.e., the vector separation of two antennas receiving a signal from a radio source. This can easily be seen by the following. Consider a right handed coordinate frame with the X axis along the Greenwich meridian and the Z axis toward the Conventional International Origin (CIO) pole.

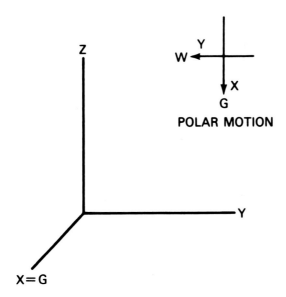

Figure 1.   Right handed coordinate reference frame with X axis towards Greenwich and Z axis towards north.

The baseline (X, Y, Z) joining the two antennas in topocentric coordinates will be rotated to

$$\begin{Bmatrix} X^1 \\ Y^1 \\ Z^1 \end{Bmatrix} = \begin{Bmatrix} 1 & +\alpha & -x \\ -\alpha & 1 & y \\ x & -y & 1 \end{Bmatrix} \begin{Bmatrix} X \\ Y \\ Z \end{Bmatrix} \qquad (1)$$

where α is the hour angle displacement of the equinox, and x and y are the polar coordinates relative to the CIO pole.

The change in baseline coordinates (ΔX, ΔY, ΔZ) may be expressed as

$$\Delta X = +\alpha Y - x Z ,$$

$$\Delta Y = -\alpha X + y Z , \qquad (2)$$

$$\Delta Z = x X - y Y .$$

One can see that changes in UT1 effect only the equatorial baseline coordinates X and Y. Elsmore (1973) pointed out that an east-west baseline is not sensitive to the effects of polar motion to first order (see equation 2) and therefore UT1 can be measured directly using a single east west baseline. To measure polar motion successfully, there must be a significant north-south component in the baseline length. With a significant north-south component in baseline length, it is impossible to measure all three components of the Earth's motion (UT1, x, and y) from a single baseline as is easily seen by inspection of equation 2. To measure all three components, it is necessary to use two independent baselines that preferably should be perpendicular. These changes in baseline coordiinates with respect to the instantaneous axis of rotation of the Earth together with the observed source's position on the sky define the interferometer phase, delay and fringe rate. For example, to first order the phase in radians is

$$\theta = 2\pi [Z \sin\delta + \cos\delta (X \cos\lambda + Y \sin\lambda)] \qquad (3)$$

where X, Y, and Z are now the baseline components in wavelengths in a right-handed coordinate frame where δ is the source declination and λ is the source right ascension less the sidereal time at Greenwich. A change in phase, $\Delta\theta$, due to the effects of Earth rotation and polar motion may be expressed as

$$\Delta\theta = 2\pi [\Delta Z \sin\delta + \cos\delta (\Delta X \sin\lambda - \Delta Y \cos\lambda)], \qquad (4)$$

or

$$\Delta\theta = 2\pi \; [+ \; \alpha \; (-X \cos \delta \cos \lambda + Y \cos \delta \sin \lambda)$$

$$+ \; x \; (X \sin \delta - Z \cos \delta \sin \lambda)$$

$$+ \; y \; (-X \sin \delta - Z \cos \delta \sin \lambda)] \; . \qquad (5)$$

Since delay is the rate of change of phase with frequency and fringe rate is the rate of change of phase with time, similar relationships may be derived for these observables. Inspection of equation (4) shows that its time derivative has no dependence on the Z component of the baseline, and is therefore sensitive to changes only in the X and Y components of the baseline. For equatorial baselines, fringe rate data, as well as phase and delay data, can yield values of UT1 from a single baseline.

The sensitivity of radio interferometric measurements of Earth rotation and polar motion is dependent on the fractional accuracy to which the baseline length may be measured in the plane perpendicular to the axis of rotation around which the motion takes place; for UT1 it is the equatorial baseline length. For example, if a 3.6 km east-west baseline is used to achieve an accuracy of 4.3 ms in UT1 measurement (Elsmore 1973) the baseline length must be measured to an accuracy of 1.1 mm, while for a 3600 km east-west baseline, the corresponding accuracy is 1.1 m.

For accuracies approaching 1 ms in UT1 and 0."01 (30 cm) in polar motion, the baseline length must be measured quite accurately. The shorter baseline techniques employing common local oscillators have baselines ranging from 3.6 to 35 km. For the longest equatorial baselines used of approximately 20 km, this accuracy corresponds to 1.4 mm. For baseline lengths used in VLBI which are approximately 3500 km this accuracy is 25 cm.

ACCURACY LIMITATIONS

There are several sources that contribute to inaccuracies in baseline measurement. The largest contributing error is caused by inaccuracies in knowledge of the differential path delay in the atmosphere of the radio signal between its arrival at the two antennas. The ionosphere will not be considered here because by making dual frequency measurements, the ionospheric delay may be calculated. It has an inverse frequency-squared dependence allowing it to be modeled accurately. For the atmosphere, the zenith path-length delay is approximately 2.3 meters for the dry atmosphere which is nearly constant, and between 1 and 25 cm at the zenith for the wet component. At 18 cm the zenith pathlength in the ionosphere and atmosphere are approximately equal but of opposite sign. The delay in the atmosphere may be modeled with ground based observations of temperature, pressure, and relative humidity. This may lead to an accuracy of $\sim$ 1-2 cm (Hopfield, 1971) in determination of the zenith

path delay at each site. Improved accuracy in measurement of the wet
term may be achieved by using microwave radiometers to measure the
column density of water vapor along the line of sight to the source.
The accuracy of this method is hoped to approach 1 cm. These measurements are important for long baselines where the atmosphere at the
antenna sites is largely independent. For shorter baselines less than
35 km, where the atmosphere may not be independent, the differential
atmospheric phase path between the antennas may be more easily obtained
through simple models. Figure 2 displays the rms phase fluctuations
converted into path-length versus baseline length for baselines less
than 35 km. These are short term fluctuations on the order of 10-15
minutes that may be averaged out with longer integration periods. Lack
of knowledge of the exact differential path length of the signal in the
atmosphere may contribute from < 1-3 mm error in the path-length for
short baselines and several centimeters for longer baselines.

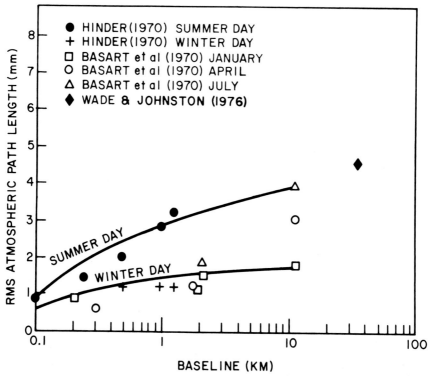

Figure 2. Variation of rms phase fluctuation versus baseline length
for short baselines.

The shorter the baseline employed, the more the observations are subject
to systematic effects that may cause small changes in baseline length.
Deformation of the antenna structure and misalignment of the antenna
antenna axis will lead to substantial systematic errors if they are not

accounted for. The reference point or phase center for an antenna depends upon its construction. For an alt-azimuth antenna, the azimuth rotation plane and elevation axes usually intersect, making this the reference point of the antenna. The rotation axes on equatorially mounted antennas usually do not intersect. The reference point is along the hour angle axis where it orthogonally intersects a plane containing the declination axis. In calculation of the correlated interferometer phase, there is a term for the lack of axis intersection. For non-intersection of axes in the alt-azimuth case the term is k cos (el) while for the equatorial case it is k cos (δ) where el is the source elevation, δ is the source declination, and k is the separation between the points where the axes should intersect. In addition to these axis offset terms, there are additional terms for axis misalignment. For an equatorial telescope the effect on phase is given by (Wade 1970)

$$\Delta\phi = -\ell \sin \delta \cos H - m \sin \delta \sin H, \qquad (6)$$

where ℓ and m are the projection of the misalignment of the x and y axes, which are perpendicular to the polar axis of the telescope, δ is the source declination and H is the source hour angle. These terms, if not solved for, will give substantial errors in source position for the Green Bank interferometer where k = 60 wavelengths at 11.1 cm or approximately 0".5 on the 2.7 km baselines and approximately 0".05 for the 35 km baseline. For equatorial antennas these terms correlate with solutions for source position and baseline depending upon the selection of observed sources and the method of observation.

Antenna movements caused by deformations of the Earth's crust must also be taken into account. Earth tides cause a substantial motion in the individual antenna positions amounting to approximately 20 cm. Again, since this is a differential effect, for baselines, less than 35 km, it will amount to less than 1 mm while for transcontinental baselines it may reach about 40 cm. The Love numbers predicting the amplitude of this effect are known to 10%, putting this effect at the many centimeter level for the longer baselines. The amplitudes of the deformation now become variables to be solved for in the baseline solution.

The structure of radio sources may put a limitation on the determination of precise positions of these objects. Comparison of precise positions of radio sources determined by short and long baselines may not be valid without correction for source brightness distribution. Short-baseline (D < 10 km) observations at 6 cm refer to radiation from spatial sizes 1 arc second or less while longer baselines (D > 3000 km) at 6 cm refer to radiation having a size structure less than 0".004. The angular distribution of radio brightness of radio galaxies, Quasars and BL Lac objects is complex with component sizes ranging from about 0".01 to well under 0".001. Quasars and BL Lac objects have high intrinsic intensity and energy density. Variations in intensity and structure occur in a time scale of weeks to months which may have to be monitored in order to compare measurements made with different spatial resolution. These

changes in structure may limit the comparison of positions made with
different spatial resolution to the 0".01 level or even the stability of
the position to this level. Thus far, at this level the positions
appear to be coincident (Wade and Johnston, 1977). It is also assumed
that since these objects are extragalactic, they will display no noticeable proper motion, i.e. 0".0001/yr. However variations in source structure over a period of years may shift the centroid of the radio brightness distribution substantially even at a single frequency and spatial
resolution.

Relativistic effects caused by motions of the Earth-Moon system and by
the gravitational field of the Sun must be accounted for and are at the
0".01 level. The values of astronomical constants which describe precession and nutation will have to be improved via radio observations.
Daily observations to an accuracy of 0".01 will improve the short period
nutation terms and as the observational time base approaches ten years,
the value of the precession constant will be improved. These effects
will be < 0".01. In summary these effects at present are at the millimeter level for short baseline lengths and five centimeter level for
VLBI baseline lengths.

REPORTED RESULTS

The results published in the literature thus far for short baselines are
those of Elsmore (1972) who used a 3.6-km baseline over a nine month
period to measure UT1 to an accuracy of 4.3 ms. For longer baselines,
Shapiro et al. (1973) reported an accuracy of 2.9 ms in UT1 measurement
with a baseline of 3899 km over a period of approximately one year.
More recent observations will be reported later in this symposium.

There is a definite need for dedicated systems to measure Earth rotation
parameters. Interested users will be applying this technique for the
measurement of Earth rotation and polar motion. Only dedicated systems
will make the daily or weekly measurements which will allow the evaluation of the true accuracy of this technique. These dedicated systems
will be the topic of presentations being made by the National Geodetic
Survey and the U.S. Naval Observatory/Naval Research Laboratory effort.

A FUTURE NEW SYSTEM

New instruments are coming on line. One instrument which may be of
interest, but is not discussed by others at this symposium, is the Very
Large Array (VLA) of the National Radio Astronomy Observatory. The VLA
in Socorro County, New Mexico, when completed in 1981, will have the
ability to measure UT1 and both components of polar motion. This instrument consists of twenty-seven, 25-meter antennas which will be located
on an equiangular "y" configuration, each arm 21 km in length with one
arm of the "y" pointing slightly off a north-south line. The three
antennas at the outermost points on the "y" will have baselines of

AN IMPROVED POLAR MOTION AND EARTH ROTATION
MONITORING SERVICE USING RADIO INTERFEROMETRY

William E. Carter, Douglas S. Robertson, Michael D. Abell
National Geodetic Survey, National Ocean Survey
National Oceanic and Atmospheric Administration
Rockville, Maryland 20852, U.S.A.

ABSTRACT

The National Geodetic Survey (NGS) of the National Ocean Survey (NOS), a component of the National Oceanic and Atmospheric Administration has begun a project to establish and operate a 3-station network of permanent observatories to monitor polar-motion and Earth rotation (UT1) by radio interferometric observations of quasars. The project designation is POLARIS (POLar-motion Analysis by Radio Interferometric Surveying).

The POLARIS observatories will be equipped with a new generation of instrumentation and software, the Mark III data acquisition and processing system currently under development by a multi-organizational team.

1. INTRODUCTION

According to plate tectonics concepts, the outer skin or crust of the Earth, called the lithosphere, consists of a dozen or so large, relatively rigid plates that float on the partially molten aesthenosphere. New lithospheric material wells to the surface from deep within the Earth along the ocean ridges. The plates tend to migrate away from these spreading centers. When they collide, large stresses and strains develop causing volcanism, earthquakes, mountain building, faulting, and the subduction of the edges of some plates.

The combined effects of the plate motions and distortions result in changes in the relative locations of some topographical features by as much as several centimeters per year. Measurements of these motions, both on local and global scales, are expected to be very useful in quantifying plate tectonics concepts. They may prove to be very useful indicators for predicting impending cataclysmic events, such as earthquakes.

Present operational geodetic methods of measurement do not have either the spatial or temporal resolution required for large scale geodynamic investigations. All of the presently known developmental techniques which are considered capable of achieving the required resolutions are astronomical or space techniques. These measurements are made in some exterior, or non-Earth-fixed, frame of reference. The transformations of the measurements to an Earth-fixed frame of reference requires knowledge of the orientation of the Earth, with respect to the non-Earth-fixed frame of reference, at the time of the measurements. Information presently available on polar motion and Earth rotation, which is largely derived from optical stellar observations, does not have sufficient spatial or temporal resolutions for use with these ultra-high accuracy methods.

The National Geodetic Survey, NOS, has reviewed the candidate methods for making geodetic measurements of sufficient accuracy for geodynamic applications. Our conclusion is that independent station radio interferometry, commonly referred to as Very Long Baseline Interferometry (VLBI), is the best method now available. This opinion is supported by the results of the series of experiments reported on by Robertson, et al. (1978), the successful completion of the initial field tests of the third generation (Mark III) VLBI data acquisition and processing systems during September 1977, and the impressive performance predicted by computer simulations for very modest VLBI networks.

We are planning to implement a comprehensive program in geodynamics which will make extensive use of radio interferometric surveying. The vanguard project in this program is known as POLARIS. Under this project, NGS will establish and operate a 3-station network of permanent observatories to determine polar motion regularly to ±10 cm and UT1 to ±0.1 milliseconds in averaging periods of less than 24 hours. This paper presents the status and plans for POLARIS.

2. THE POLARIS NETWORK

NOAA is working closely with the National Aeronautical and Space Administration (NASA) to establish the POLARIS network. Two existing, lightly used, radio telescope facilities have been identified for probable use in project POLARIS: the 18-meter diameter Westford telescope at the Northeast Radio Observatory Corporation site near Boston, Massachusetts, and the 26-meter diameter telescope at the Harvard Radio Astronomy Station (HRAS) near Fort Davis, Texas.

The Westford facility is particularly attractive because it is located less than 1.25 kilometers from the Haystack Observatory, which originated much of the design and developmental work on the Mark III

VLBI system. NGS will be able to rely on the expertise of the Haystack staff and time-share the Haystack multi-base line Mark III correlator. Other advantages include the joint use or comparison of local oscillators via the actively stabilized land line which connects the two facilities, joint observing programs to ensure the proper functioning and calibration of both systems, joint system improvement developmental programs, and cooperative data analysis.

The HRAS site also has some unique attributes. It is located less than 8 kilometers from the University of Texas McDonald Observatory, which has been used regularly since 1969 for lunar-laser ranging experiments (LURE). NASA, the University of Texas, and NOAA already have a cooperative program to monitor the stability of a region extending more than 100 kilometers from the McDonald and HRAS sites. A broad mixture of techniques are being used. These include electromagnetic distance measurements, spirit levelling, gravimetry, seismometry, and tilt measurements (Carter and Vincenty, 1978; Dorman and Latham, 1976).

Other existing facilities currently under consideration are the National Environmental Satellite Service Observatory near Fairbanks, Alaska, and a facility at the NASA Goldstone complex near Barstow, California.

Serious consideration is also being given to the construction of a new radio observatory on the grounds of the U. S. Naval Observatory's (USNO) Time Service substation, near Richmond, Florida. A network including the Westford and Richmond sites, which are both very near the Atlantic Coast, would be excellent for cooperative programs, including both geodynamic and time transfer, with European observatories. Shared operations between NGS and USNO would be economically desirable for both agencies.

The selection of the POLARIS sites will be completed during 1978, and instrumenting of the first two observatories will begin immediately thereafter. Cooperative observing sessions with other observatories which are being instrumented with Mark III VLBI systems may begin as early as July 1979. The goal is to have the total POLARIS system fully operational by the end of 1981.

3. INSTRUMENTATION

The POLARIS network will use the Mark III VLBI data acquisition and processing system. A simplified block diagram of this system is shown in Figure 1. Briefly, the more important characteristics include:

- Very wide band receiver front end (400 MHz) for high delay resolution.

- Dual frequency (X and S band) receiving systems for extraction of the ionosphere delay.

- Continuous system calibration.

- Very stable hydrogen maser clock system.

- Water vapor radiometer and meteorological sensors for the determination of the tropospheric delay.

- A wide bandwidth (56 MHz) data recording system for good signal-to-noise ratio, even with small (10-meter diameter) antennas.

- Automated control and monitoring with a mini-computer in order to facilitate easy field operations.

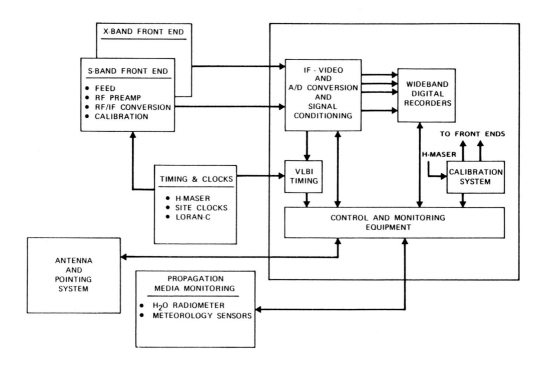

Figure 1. Block Diagram of Mark III VLBI Data Acquisition System.

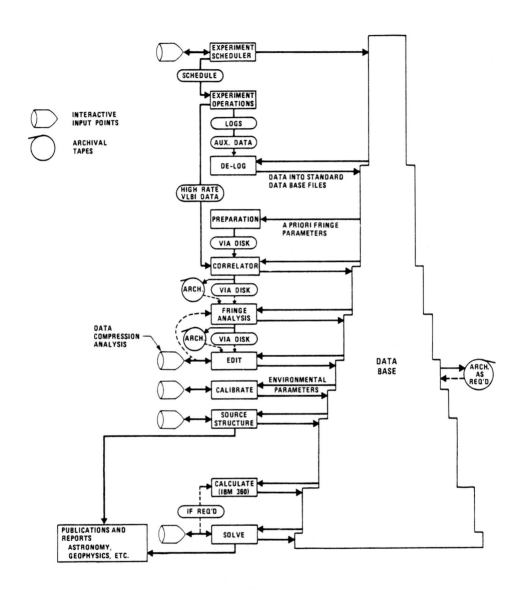

Figure 2. Schematic of the Mark III VLBI Data Processing System.

A completely new set of well-documented software has been developed as an integral part of the Mark III VLBI system. Figure 2 is a schematic representation of the system. The most salient characteristic is the structuring of the various programs around a central data base. A variety of programs, which accomplish such diverse functions as generating observing schedules, processing data, and preparing publications, all interact with this single data base.

The primary programs, i.e., those used to do the actual reduction of VLBI data to extract the geodetic information, are operational. In fact, the data reduction and analyses for the paper by Robertson, et al. (1978) were performed with the Mark III software but using Mark I data, on a mini-computer system at NASA Goddard Space Flight Center.

4. CONCLUDING REMARKS

The expected spatial and temporal resolutions of the POLARIS data constitute at least an order-of-magnitude improvement over presently available data. Until data of this quality have actually been collected and analyzed, we will not know the most desirable operating mode. The timely analysis of data and the adjustment of procedures and schedules for responding to the newly acquired knowledge from the analyses will be particularly important during the first few years of POLARIS operations.

The National Geodetic Survey intends to seek actively the advice and cooperation of the international scientific community. The POLARIS data will be disseminated as quickly as possible in the most convenient form that can be devised within budgetary constraints.

It is hoped that other nations will give serious consideration to establishing one or more observatories suitably instrumented and staffed for operational use of radio interferometry. We look forward to the development of an international network of observatories working cooperatively to advance both the theoretical and practical aspects of the earth sciences.

REFERENCES

Carter, W. E. and Vincenty, T.: 1978, "Survey of the McDonald Observatory Radial Line Scheme by Relative Lateration Techniques" NOAA Technical Report NOS 74 NGS 9.

Dorman, J. H. and Latham, G. V.: 1976, "Preliminary Geophysical and Geological Site Survey of the Region of the McDonald Observatory, West Texas" Final Technical Report, NASA Contract NGS 7159.

Robertson, D. S., Carter, W. E., Corey, B. E., Counselman, C. C., Shapiro, I. I., Wittels, J. J., Hinteregger, H. F., Knight, C. A., Rogers, A. E. E., Whitney, A. R., Ryan, J. W., Clark, T. A., Coates, R. J., Ma, C. and Moran, J. M.: 1979, this volume.

DISCUSSION

I. I. Mueller: Will the stations be available for 24 hours per day?
W. E. Carter: Fort Davis belongs to Harvard College, but we hope to get 50% of each day; Westford is available almost full time and if we build another station in Florida that probably will be, too.
K. Johnston: Is Fort Davis a reliable VLBI site, and does the Fort Davis - Haystack baseline really give optimum geometry?
W. E. Carter: We intend to update the Fort Davis antenna; Fort Davis is a good location within the USA and is close to other sites where relevant work is done.

DETERMINATION OF UT1 AND POLAR MOTION BY THE DEEP SPACE NETWORK USING VERY LONG BASELINE INTERFEROMETRY

J. L. Fanselow
J. B. Thomas
E. J. Cohen
P. F. MacDoran
W. G. Melbourne

B. D. Mulhall
G. H. Purcell
D. H. Rogstad
L. J. Skjerve
D. J. Spitzmesser

California Institute of Technology
Jet Propulsion Laboratory, Pasadena, California

Jose Urech
Institute National De Technica Aerospacial
Madrid, Spain

George Nicholson
National Institute for Telecommunications Research
Johannesburg, Republic of South Africa

ABSTRACT

The Deep Space Network (DSN) [operated by JPL under contract to the National Aeronautics and Space Administration] is implementing a Very Long Baseline Interferometry (VLBI) capability at DSS 63 (Spain), DSS 14 (California, USA), and DSS 43 (Australia) to support the navigation requirements of planetary space missions. The early development work for this system has already demonstrated the capability of measuring UT1 with a formal accuracy as low as 0.6 msec with only 6 hours of data. Further, a radio astrometric catalog of approximately 45 sources whose positions are known to better than $0\overset{\prime\prime}{.}05$ has been constructed. In addition to these measurements, this paper describes the characteristics and anticipated performance of the complete VLBI system being implemented within the DSN for operational use in mid-1979. In particular, one of the capabilities of this system will be the measurement of UT1 and polar motion at weekly intervals. Although the navigation accuracy requirement is only 50 cm for the Voyager mission, this system should be capable of delivering UT1 and polar motion determinations with decimeter accuracy if it is operated at maximum performance. An additional requirement of this operational system is that it have the capability of providing these results within 24 hours of the actual observations.

INTRODUCTION

Independent station radio interferometry, more commonly known as VLBI (Very Long Baseline Interferometry), is a technique that was pioneered in 1967 by radio astronomers to study the structure of compact natural radio sources. Since then various groups throughout the world have expanded the applications of VLBI to geodesy, spacecraft navigation, astrometry, and to clock synchronization. In this paper we will discuss the work of only one such team, that of Caltech's Jet Propulsion Laboratory (JPL). Furthermore, we will discuss only the effort directly connected with the Deep Space Network (DSN).

DESCRIPTION OF THE EXPERIMENTS

Since 1971, the Jet Propulsion Laboratory of the California Institute of Technology has been developing VLBI for application to geodesy and spacecraft navigation under contract to the National Aeronautics and Space Administration. We will discuss only the work associated with the Deep Space Network. The Deep Space Network consists of a ground communications facility, a network operations control center at JPL, and three complexes of very low noise radio antenna systems. One complex is in Spain, near Madrid, another is at Goldstone in southern California, U.S.A. and the third complex is near Canberra, Australia. Intercomplex baselines of 8-10 thousand kilometers are thus available. At each complex there are at least two 26-meter diameter antennas, and one 64-meter diameter antenna. Typical system temperatures of 25-30°K at both S Band ($\lambda \sim 13$ cm) and X Band ($\lambda \sim 3.8$ cm) are available.

Table 1 summarizes the observations which have produced the results in this report. Up to this time, the emphasis of the program has been system development, including hardware and software. No attempt has been made to routinely collect data at times optimized to improve geophysical understanding of the Earth's rotation.

In the results to be presented, all of the data summarized in Table 1 were simulatneously fit in a least-squares type estimation process. Both delay and delay rate were fit when both were measured. This produced a total of about 1000 observables with which were associated approximately 270 estimated parameters. The delay observable was usually dominant where both delay and delay rate were included. We chose to adjust only those model parameters whose a priori determinations were less accurate than the sensitivity of our data to that parameter.

The following describes the model in more detail. In the geometric portion of the delay model we estimated radio source positions, station locations, and UT1-UTC. We did not estimate polar motion, the parameters of the solid Earth tide model we used, nor the precession and nutation constants.

The model for delay was calculated employing special relativity to all

Table 1. Preliminary results for UT1 from 1971-77 VLBI data.

DATE	BASELINE°	NO. OF OBS	UT1 VLBI-BIH (msec)	DATE	BASELINE°	NO. OF OBS	UT1 VLBI-BIH (msec)
8/28/71	14/62	72	2.9±2.3	1/12/77	11/43	12	1.1±2.3
9/1/71	14/62	24	1.1±2.2	1/21/77	14/43	28	-2.8±1.6
9/1/71	51/62	22		1/21/77	11/43	22	
9/6/71	14/62	45	-2.4±2.0	1/31/77	14/63	27	REFERENCE
9/10/71	14/62	45	-1.3±2.4	2/1/77	11/43	24	REFERENCE
4/30/73	14/62	24	-3.1±2.8	2/13/77	14/43	40	-2.3±1.6
9/8/73	14/62	17	5.1±3.5	2/13/77	11/43	34	
11/20/73	51/63	12	21±10	2/28/77	14/43	64	-5.3±1.7
2/15/74	14/62	20	1.6±2.7	2/28/77	11/43	9	
4/21/74	14/62	22	3.4±3.0	4/13/77	14/63	48	-0.1±0.6
6/21/74	14/62	17	4.2±4.5				
8/6/74	14/62	7	3.6±4.4				

°DSS 11, 14 IN CALIF.; DSS 51 IN S. AFRICA; DSS 62, 63 IN SPAIN; DSS 43 IN AUSTRALIA

orders of v/c, and using the Earth's velocity about the center of mass of the solar system. Diurnal polar motion, the effects of a liquid core, and general relativity were not included. However, the effects of these deficiencies in our model are thought to be less than our data noise with the current data.

Our a priori model used the Bureau International de l'Heure (BIH) Circular D smoothed values for UT1-UTC, and polar motion. A priori station locations were those that have been obtained over the years by spacecraft tracking. If a radio source had an optical counterpart which could be placed on the FK-4 system, we used that position as a priori, and included its coordinates with the proper statistical weight (typically a few tenths of an arc second). Otherwise, we used positions obtained by single radio antenna measurements.

Although VLBI can determine source declination absolutely, a coordinate origin definition on the celestial sphere must be made for right ascension. Likewise, a baseline longitude (or alternatively, UT1) definition for the Earth must be made. In an effort to mesh our results with the conventional coordinate systems, we have made the following definitions.

1. The right ascension of the radio source NRAO 140 was fixed at a given value in the final fit. This particular right ascension value was determined by means of a preliminary fit in which the right ascensions of all the optical counterparts were statistically constrained by their a priori error. This defined the right ascension origin and produced an overall shift in the right ascensions of all the sources such

that the offset between the radio reference frame and the FK-4 system was approximately 0."1 or less.

2. The Earth-fixed longitude origin for the baselines was established by adopting the BIH values for UT1-UTC on January 31, and February 1, 1977 as exact. (Since two main baselines were used, and the data on these baselines did not overlap, fixing the value of UT1-UTC on two days was required).

With regard to the clock contribution, the effect on the observed time delay of the different epochs and rates of the two independent oscillators for each station pair was modeled as a function linear in time. In the least-squares estimation process the parameters representing the differential epoch and rate were given independent status for each experiment. On several days, however, it was necessary to break the clock model into several time segments within the day, fitting a separate linear function within each of the segments.

We modeled the troposphere as a spherical shell with the thickness specified as a parameter to be estimated. For each experiment, a new estimate of this thickness was allowed for each station. For each station, and for each date, the a priori troposphere thickness was obtained from a table of monthly mean values developed for the DSN by C. Chao. The a priori error of these values, which was used to constrain the estimated troposphere parameter, turned out to be roughly equal to the error which the VLBI data alone would have produced in estimating the troposphere thickness.

For the ionosphere we used a simple model of the total electron content as a function of local time. This model was only an approximation. However, more detailed models made little difference. Because we plan to remove the effects of the ionosphere by simultaneously observing at both S and X Band, no significant effort is being directed to improve this model.

RESULTS

With the models and constraints defined above, a simultaneous multi-parameter fit was made to all experiments to obtain estimates of the parameters. Figure 1 plots the UT1 (VLBI) - UT1 (BIH) values given in Table 1. Notice the improvement that results from measuring delay in addition to delay rate. Note also that the most recent point has a formal uncertainty (1σ) of 0.6 msec with about 6 hours of data. This last observing session was on the nearly east-west Goldstone-Spain baseline. The geometry of that baseline lends itself very well to the determination of UT1 at the same time source positions are being estimated.

Baseline vectors are closely tied with UT1 measurements. Table 2 presents the various components of the baselines relevant to the

DETERMINATION OF UT1 AND POLAR MOTION

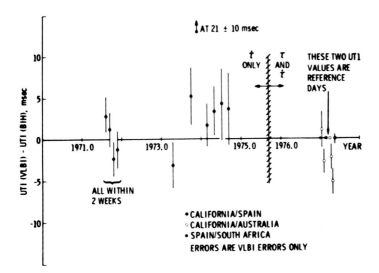

Figure 1. UT results for 1971-77 VLBI data.

Table 2. Preliminary results from intercontinental baselines from 1971-77 VLBI data.

BASELINE	CALIFORNIA/SPAIN (14/63)	CALIFORNIA/ AUSTRALIA (14/43)	SPAIN/SOUTH AFRICA (51/63)
EQUATORIAL LENGTH (m)	8378987.2 ± 0.3	7620842.95 ± 0.45	3037637.5 ± 0.9
"LONGITUDE"	$30°.726453 \pm 1.5$ msec (1 m)	$106°.052285 \pm 2$ msec (1.2 m)	$265°.537323 \pm 6$ msec (1.3m)
POLAR COMPONENT(m)	438056.1 ± 1.2	-7351802.3 ± 1.3	
TOTAL LENGTH (m)	8390430.23 ± 0.30	10588968.0 ± 1.0	

experiments we are reporting. The equatorial length and total length of the California-Spain baseline was determined with a formal uncertainty (1σ) of 30 cm, while the polar component uncertainty was 1.2m. The uncertainties on the components of the other two baselines ranged between 0.4 and 1.3m. Notice a typical problem with long baselines, the relative difficulty in measuring the polar component. This is because that parameter tends to correlate with other parameters. Also, note that because we did not have measurements of time delay on the Spain-South Africa baseline, we are unable to measure that baseline's polar component.

To eventually be able to measure Earth orientation with arbitrary blocks of antenna time, it is desirable to have the sky well covered with sources whose positions are well known. Figure 2 schematically displays the sky coverage of our current source position catalog. In the interest of clarity, we have plotted only those sources for which the error in right ascension or declination (or both) was $\leq 0\rlap{.}''1$. Note that the scale for the errors is different than the scale for positions.

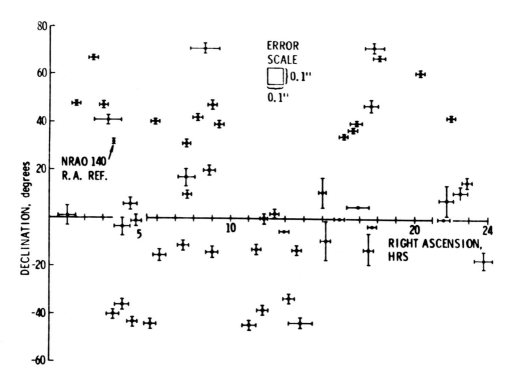

Figure 2. Distribution of sources from 1971-77 VLBI data. Only sources with an error $\leq 0\rlap{.}''1$ are included (S-band only).

DETERMINATION OF UT1 AND POLAR MOTION

Also the lack of an error bar implies that the error in that component was $\gtrsim 0\overset{\prime\prime}{.}1$. Only about one half of the sources in our eventual catalog have had their positions measured with an accuracy better than $0\overset{\prime\prime}{.}1$.

All these data are preliminary and are for S-band only. Inclusion of X-band data, and the solving for relevant components of polar motion must be performed before we consider the analysis of these data completed. However, even these preliminary results indicate a current capability of measuring source positions and Earth orientation with a formal accuracy of $0\overset{\prime\prime}{.}01 - 0\overset{\prime\prime}{.}04$ and intercontinental baselines at a 0.3-1.5m level.

ANTICIPATED VLBI CAPABILITY OF THE DSN FOR EARTH ORIENTATION MEASUREMENTS

The current capability discussed above has been demonstrated during engineering development tests for an operational VLBI system. This system is being implemented within the Deep Space Network for the support of spacecraft navigation and will be operational by July, 1979. The remainder of this paper outlines the characteristics of this particular system, and the projected VLBI capabilities of the DSN into the early 1980's.

Table 3 summarizes the capabilities of the VLBI system being implemented. Its primary purpose is to monitor the hydrogen maser oscillators within the network and thereby determine if their stability meets the oscillator requirements of the Voyager mission to Saturn. In this role the system will be used on a weekly basis to determine clock offsets between the three DSN complexes. In the analysis of the weekly measurements, a priori radio source positions amd Earth fixed station locations will be provided by other longer, but less frequent, VLBI measurements. Data will be taken for approximately two hours on the California-Spain baseline, followed within 24 hours by another two hours of data on the California-Australia baseline. For each of the two sessions, the clock epoch offset and rate offset will be estimated. These clock parameters, along with UT1 and both components of polar motion, will be simultaneously estimated in a fit to the data from both observing sessions. Thus, each pair of 2-hour sessions will yield estimates of 7 parameters - three Earth orientation parameters and four clock parameters.

The required accuracy is: 10 nsec for clock epoch offset, 3×10^{-13} for fractional frequency offset, and 50 cm for Earth orientation. Furthermore, there must be the capability to have these results within 24 hours of the time the data is taken. These specifications refer to a "worst-case" performance of the system. It is a capability that must exist for arbitrary assignments of 2-hour antenna time blocks. In actual practice the system performance can be somewhat better. Specifically, we expect to determine true clock epoch offsets to 2-4 nanosec (limited by calibrations of cable lengths). Furthermore, since uncertainty in a priori source positions will initially be a dominant error source, the expected steady improvement of the source position catalog wil lead to increasingly better Earth orientation values. Since improved positions can be applied to earlier data at any time, significant revisions of earlier real time results should be possible as new position data are incorporated a year or two later.

Table 3. Characteristics of DSN operational VLBI system.

Antenna Diameters	64 meters
Antenna Locations	Spain (DSS 63) USA (DSS 14) Australia (DSS 43)
Observing Frequency	S-band - 2.3 x 10^9 Hz simultaneous X-band - 8.4 x 10^9 Hz
Record Rate	500 Kbs (4 Mbs if not in near real time mode)
Spanned Bandwidth	40 MHz, S-band 50 MHz, X-band
Frequency Standards	Hydrogen Masers ($\Delta f/f \lesssim 10^{-14}$)
Nominal Interval between Observation Pairs (DSS 14/63, DSS 14/43	1 Week
Nominal Observation Time on each Baseline	2 Hours
Nominal Accuracy	
UT1, Polar Motion	30 cm
Fractional Frequency Offset	10^{-13}
Clock Epoch Offset	<10 nsec
Lag between Data Acquisition and Results	<24 hours

Special observing strategies also are feasible where increased accuracy of the variations in Earth orientation are desired. Longer, and more frequent, observing sessions, along with observing sessions chosen to utilize a particular portion of the source catalog, would greatly improve the accuracy with which variations in UT1 and polar motion could be measured. Use of the operational system to obtain short term variation measurements of accuracy $\sim 0\rlap{.}''003$ on a daily, or twice daily basis, is possible by late 1979, or early 1980. However, with the anticipated resource allocations, routine weekly measurements of $\sim 0\rlap{.}''01$ accuracy are expected.

In support of this operational VLBI system, more advanced VLBI systems will be employed within the DSN on a development basis. These systems will be used to refine the radio source catalog and the relative Earth fixed station locations, as well as to improve the modeling capability of the software, and to test new hardware. Consequently, not only will there be a steady improvement in the accuracy of the operational system, but there will also be a "best efforts" capability in the advanced VLBI system far superior to that of the operational system.

Figure 3 schematically illustrates the anticipated accuracy of this development effort as a function of time. The solid line shows the purely instrumental errors as new hardware becomes available, while the dotted line is the overall accuracy for measuring Earth orientation. Note, the delivered performance always lags well behind the instrument capability. Such a lag results from software development with Earth modeling capability always lagging the instrument development. However, as Figure 3 indicates, we anticipate a $0\rlap{.}''002$-$0\rlap{.}''003$ capability in the early 1980's with 12 hours of data on each of the two baselines, California/Spain, and California/Australia. This accuracy would not be limited to short term variation measurements, but would apply to all uses of the system for Earth orientation determination.

Certain assumptions, of course, have been made. One is that the positions of natural radio sources are time invariant. Some experiments by other groups have already indicated that this is a valid assumption. Experiments are now underway which will test this assumption more thoroughly. Another assumption is that integrating the Earth's motion for 12 hours is meaningful at the $0\rlap{.}''002$ level. Again, that will have to be determined experimentally.

There are also other problems which must be considered before measurements at this accuracy are useful. For example, the current definitions of UT1 and polar motion are not strictly correct once measurement accuracy reaches levels comparable to the yearly movement of the crustal plates of the Earth. Another problem concerns the best definition of a right ascension origin for the radio source catalogue at the $0\rlap{.}''001$ level. Practical considerations are going to force us into making working definitions for our use of VLBI by 1981. There is clearly a growing need for the consultation and cooperation of the international community in defining coordinate systems suitable for milliarcsecond measurements of Earth dynamics and source catalogues.

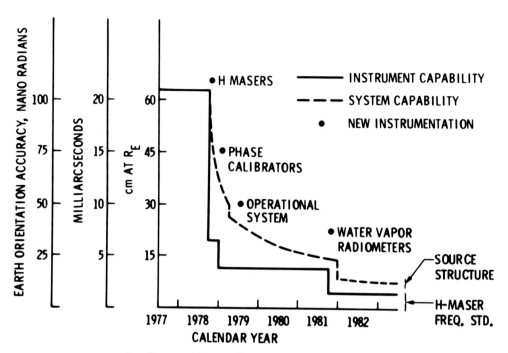

Figure 3. "Best efforts" capability for determining Earth orientation with 12 hours of VLBI data on each of 2 baselines: DSS 14/43 and DDS 14/63.

ACKNOWLEDGMENT

This paper presents the results of one phase of research carried out at the Jet Propulsion Laboratory, California Institute of Technology, under Contract No. NAS 7-100, sponsored by the National Aeronautics and Space Administration.

DISCUSSION

J. D. Mulholland: The need for the source to be visible at both sites means that 12-hour integrations are not possible on a single source. The predictions assume that more than one source will be observed during a run.

J. L. Fanselow: We do not necessarily use a single source. Many different sources are observed during a single experiment. We generally try to observe a given source at least three times during its passage through the common visibility pattern of the interferometer. This takes about four hours, almost independent of declination, for the Goldstone-Australia baseline; for Goldstone-Spain it varies from a few minutes for declination $-20°$ to 24 hours for polar sources. We observe the polar sources many times during an experiment. The

Goldstone-Australia baseline is not good by itself but is very good in combination with Goldstone-Spain.

K. Johnston: The declination measurements are not absolute but are relative to the instantaneous spin axis of the Earth.
What is the error from a six-hour measurement of UT1?

J. L. Fanselow: I agree with your comment.
The formal 1σ error of our most recent UT1-UTC value was 0.6 ms from six hours of data. The Goldstone-Spain baseline is very suitable for the simultaneous determination of source positions and UT1-UTC.

THE NAVOBSY/NRL PROGRAM FOR THE DETERMINATION OF EARTH ROTATION AND POLAR MOTION

K. J. Johnston, J. H. Spencer, C. H. Mayer
E. O. Hulburt Center for Space Research
Naval Research Laboratory, Washington, D. C. 20375
W. J. Klepczynski, G. Kaplan, D. D. McCarthy, G. Westerhout
U. S. Naval Observatory, Washington, D. C. 20390

The joint program of NAVOBSY/NRL is discussed.

INTRODUCTION

The United States Naval Observatory (NAVOBSY) and the Naval Research Laboratory (NRL) are collaborating in a program to apply radio interferometric techniques to the determination of variations in Earth rotation, polar motion, and improved astronomical position reference systems. Investigations of VLBI and connected interferometer techniques and radio sources for astrometic application have been in progress for several years as part of the NRL radio astronomy program, and currently NRL and NAVOBSY are carrying out experimental programs to investigate VLBI time transfer techniques and UT determination using the connected element interferometer of the NRAO in Green Bank. Some previous results of observations using the Green Bank interferometer and proposed plans for operation as a dedicated system over a period of time to evaluate effectiveness for precise determination of Earth rotation parameters are discussed.

The Green Bank interferometer of the National Radio Astronomy Observatory (NRAO) includes three 26-meter antennas which are located along an azimuth of 242 degrees. Two of the antennas are moveable giving 16 possible antenna separations with the maximum separation being 2.7 km (Hogg, MacDonald, Conway, and Wade, 1969). The operating frequencies are 2695 and 8085 MHz. In addition to these three on-site antennas, a 14-meter antenna is located 35 km away. The signal from this antenna is transmitted to the site via a phase-stable radio link where it is combined with the signals from the three 26-meter antennas giving three long baselines which range in length from 33091 to 35266 m and in azimuth from 204° to 207°. The baseline geometry is shown in Figure 1.

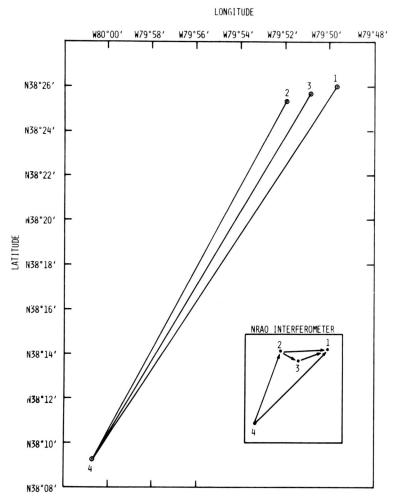

Figure 1. Geometry of the Green Bank interferometer. The three on-site 26 m antennas are designated as 1, 2, and 3 while the 14 m remote antenna is designated as 4.

Observations made with this instrument between December 1974 and January 1976 have resulted in the determination of the positions of 36 radio sources to an accuracy of a few hundreths of an arc second (Wade and Johnston, 1977). As a by-product of these observations, observations were made during January 1976 of the radio sources NRAO 140, 3C345, and NRAO 512 for the purpose of evaluating the accuracy of this instrument for the determination of Earth rotation parameters. Earlier Elsmore (1973) had demonstrated that a 3.6 km east-west baseline could determine UT1 to an accuracy of 4.6 ms. Since the three Green Bank baselines are close to parallel, both UT1 and the two components of polar motion could not be solved for independently (see Johnston, 1979). Polar motion from the Dahlgren Polar Motion Service was used to solve for UT1. Figure 2 displays the difference between the BIH value of UT1 and the value found from the radio observations. The average difference is 2.1 ± 1.2 ms.

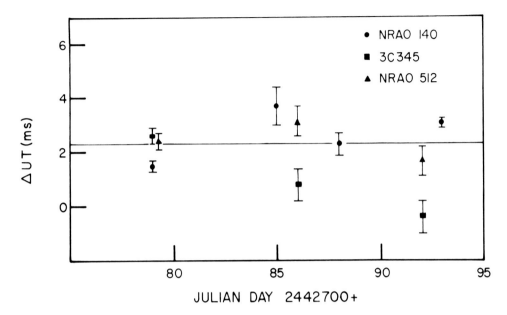

Figure 2. Measurements of UT1 made in January 76 using the Green Bank interferometer. The error bars display the rms accuracy of the solution for each determination.

There are the following advantages in using a connected element interferometer to determine Earth rotation parameters (Johnston, 1974):

1. Single station operation leading to lower cost.
2. Simple electronics - no video tape recorders and no need for hydrogen maser frequency standards.
3. Real time capability - the data can be reduced immediately allowing the immediate determination of UT1 and polar motion.
4. The system can be easily automated.
5. The relatively short interferometer baselines which are adequate with this technique minimize the effects of the atmosphere, ionosphere, and Earth tides.

However there are some disadvantages:

1. The baselines must be determined to millimeter accuracy. Deformations of the antennas may lead to significant systematic effects.
2. For the existing interferometer in Green Bank the remote 14 meter antenna is more than 200 m higher than the on-site 26 meter antennas. Variations in atmospheric conditions between the remote and main sites may lead to errors in baseline determination.

With the construction of the VLA, the National Radio Astronomy Observatory is phasing out operation of the Green Bank interferometer by October 1978. It is proposed that this facility be operated initially to evaluate the accuracy with which this instrument can determine Earth rotation parameters such as UT1 and polar motion in all weather conditions over periods of observation of eight hours duration. If the results are favorable, another long baseline can be added by locating an antenna at a second remote site to form orthogonal baselines. With two perpendicular baselines, UT1 and both components of polar motion can be found independently from the observations making this the only observatory that can precisely determine all three components of the motion of the Earth's crust.

This facility will be the first dedicated to radio astrometry. As such it will provide daily observations for the determination of Earth rotation parameters. In addition to the determination of Earth rotation parameters, this program will improve the positions of the approximately twenty radio sources observed to $\leq 0''.01$. These positions will define an almost inertial reference frame. The accuracy of the short period nutation terms will also be improved through the continuous radio observations.

REFERENCES

Elsmore,B.:1973, Nature 144, pp. 423-424.
Hogg,D.E., MacDonald,G.H., Conway,R.G., and Wade,C.M.:1969, Astrophys. J. 74, p. 1206.
Johnston,K.J.:1974, Proc. 6th Annual PTTI Meeting, Dec. 3-5, pp. 373-379.
Johnston,K.J.:1979, this volume.
Wade,C.M., and Johnston,K.J.:1977, Astron. J. 82, pp. 791-795.

DISCUSSION

E. Silverberg: Would any VLBI investigator comment on the "all weather capability" of VLBI, with particular reference to the determination of baselines to 5cm accuracy?

J. L. Fanselow: Over the past five years we have had several hundred VLBI experiments, all of them scheduled well ahead, so that we had to accept the weather that came. I am not aware of more than one or two experiments during that interval that were lost because of weather. In several experiments we had heavy rain at all sites; this increased system noise temperatures, and so reduced signal-to-noise ratios. High winds, of more than 45 miles per hour, have forced curtailment of several experiments in order to prevent mechanical damage to the antenna structure.

Observations of time delay and its rate of change, made with long baselines, are relatively insensitive to small changes in the atmosphere; for example, a 13 cm change due to the atmosphere causes only a $0''.003$ error in the typically measured VLBI angle when time delay is observed on a 10 000 km baseline.

Ya. S. Yatskiv: Your slide mentioned the determination of satellite positions; please comment.

K. Johnston: We are studying the accuracy of both conventional interferometry and VLBI for the determination of satellite positions relative to an almost inertial frame defined by celestial radio sources. At present I estimate the accuracy of the relative positions of a satellite-quasar pair as $0\!''\!.001$.

L. V. Morrison: You said that you have to determine the baseline with millimeter accuracy. Is there correlation here with UT1 and polar motion, the quantities you are trying to measure?

K. Johnston: Not really; we have used data from three one-week observing runs, corrected for nutation and similar effects, and we solved to get source positions and baselines as nearly independent as we could. There is a small problem associated with the use of equatorial mounts, because there is correlation between the determination of axis separation and source declinations, but we believe we have now overcome this.

RECENT RESULTS OF RADIO INTERFEROMETRIC DETERMINATIONS OF A TRANSCONTINENTAL BASELINE, POLAR MOTION, AND EARTH ROTATION

D. S. Robertson, W. E. Carter
National Oceanic and Atmospheric Administration
National Ocean Survey/National Geodetic Survey

B. E. Corey, W. D. Cotton, C. C. Counselman, I. I. Shapiro, J. J. Wittels
Department of Earth and Planetary Sciences and Department of Physics
Massachusetts Institute of Technology

H. F. Hinteregger, C. A. Knight, A. E. E. Rogers, A. R. Whitney
Haystack Observatory

J. W. Ryan, T. A. Clark, R. J. Coates
National Aeronautics and Space Administration

C. Ma
University of Maryland

J. M. Moran
Smithsonian Astrophysical Observatory

ABSTRACT

Radio interferometric observations of extragalactic radio sources have been made with antennas at the Haystack Observatory in Massachusetts and the Owens Valley Radio Observatory in California during fourteen separate experiments distributed between September 1976 and May 1978. The components of the baseline vector and the coordinates of the sources were estimated from the data from each experiment separately. The root-weighted-mean-square scatter about the weighted mean ("repeatability") of the estimates of the length of the 3900 km baseline was approximately 7 cm, and of the source coordinates, approximately 0.''015 or less, except for the declinations of low-declination sources. With the source coordinates all held fixed at the best available, a posteriori, values, and the analyses repeated for each experiment, the repeatability obtained for the estimate of baseline length was 4 cm. From analyses of the data from several experiments simultaneously, estimates were obtained of changes in the x component of pole position and in the Earth's rotation (UT1). Comparison with the corresponding results obtained by the Bureau

International de l'Heure (BIH) discloses systematic differences. In particular, the trends in the radio interferometric determinations of the changes in pole position agree more closely with those from the International Polar Motion Service (IPMS) and from the Doppler observations of satellites than with those from the BIH.

For the past several years, our group has been involved in a major effort to improve the quality of data obtained by very-long-baseline interferometry (VLBI) through development of a third generation system referred to as the Mark III system. Portions of this system are presently in operation, and we expect the system to be in full operation by the end of 1979. The major improvements in the hardware and software already implemented can be cataloged as follows:

First: the use of wide-band parametric amplifiers which allow bandwidths of about 300 MHz to be synthesized, a tenfold increase over the previous limit. The significance of this increase in bandwidth lies in the fact that the uncertainty in the measurement of interferometric group delay is inversely proportional to the total bandwidth.

Second: the use of surface meteorological measurements with algorithms supplied by J. Marini of NASA's Goddard Space Flight Center. This combination has resulted in a significant decrease in the level of systematic errors caused by atmospheric effects.

Third: the use of equipment to calibrate instrumental effects, in particular the effect of the contribution to the measured delay of the path from the antenna feed horn to the recording apparatus. The uncertainty in such instrumental effects has thereby been reduced to the level of a few pico-seconds.

Fourth: the inclusion of several geophysical effects, resulting from the nonrigidity of the Earth, in the analysis programs (Woolard, 1959; Melchior, 1971; Guinot, 1970, 1974). These effects have an amplitude of a few hundredths of an arc second or less. Also we employed the value of the precession constant from the paper by Lieske et al. (1977).

Fifth: the use of rewritten analysis programs. The inter-comparison of the new programs with the old indicates that there are no coding errors in the parts that were in common with effects on the interpretation of the observations larger than 1 mm. (This test does not address possible errors in the physical models on which both programs were based or in the additions noted above. For example, the precession-nutation algorithms employed are unlikely to be accurate at the millimeter level.)

Utilizing the above mentioned improvements in hardware, we undertook a series of VLBI experiments, starting in September 1976. These experiments utilized the 37-meter-diameter antenna at the Haystack Observatory in Massachusetts, and the 40-meter-diameter antenna at the Owens Valley

Radio Observatory in California, to observe extragalactic radio sources. We have used the data from these experiments to estimate the components of the baseline vector, the coordinates of the radio sources, and the changes in the x component of the position of the Earth's pole and in the Earth's rotation (UT1). (This particular baseline is not sensitive to changes in the y component of the position of the pole because any change in that component causes a nearly parallel displacement of the baseline vector and, therefore, has little effect on observations of very distant sources. In any event, a single baseline can yield information on only two of the three angles required to specify the orientation of the Earth in inertial space.)

We discuss here the observations made with this interferometer in fourteen experiments, the last of these being in May 1978. Each experiment spanned from 15 to 48 hours in which from 120 to 240 separate, three-minute, observations were made. The data from each session were first analyzed separately, and the following parameters were estimated: the vector components of the baseline, the epoch and rate offsets of the clock at Owens Valley relative to those offsets of the clock at Haystack, the zenith electrical path length of the atmosphere at each site, and the right ascension and declination of each source, with one right ascension held fixed to define the origin of right ascension. To test the consistency of the results, the repeatability of the estimates of baseline length was examined. Baseline lengths were selected for examination because the values of the direction components of the baseline vector are affected by errors in the values used for the pole position and for UT1; similarly, the estimated values for the source coordinates are affected by errors in the formulas used for precession and nutation, although the arc lengths between sources are free from such errors.

The estimate of the baseline length from each experiment is shown in Figure 1. The error bars represent formal standard errors based on measurement errors modified so that the root-weighted-mean-square scatter about the weighted mean (hereinafter "RMS scatter" or "repeatability") of the postfit residuals is unity. The RMS scatter of these baseline length values is 7 cm, or, expressed as a fraction of the 3900 km baseline, about 2 parts in 10^8. The values of the source coordinates from these solutions have an RMS scatter of 0$''$015 or less, except for the declinations of the low-declination sources for which our observations have less sensitivity. The values for these coordinates appear to be somewhat more accurate than our previously published results (Clark et al., 1976) and will be discussed more fully in a separate paper.

To examine how the repeatability of the baseline results might have been affected by the prior availability of sufficiently accurate values of the source coordinates, we obtained new solutions with each source coordinate fixed at the weighted mean of its values from the 14 separate solutions, thereby reducing the number of parameters estimated in each solution from about 30 to about 10. We would expect the RMS

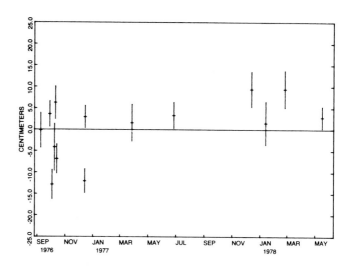

Figure 1. Differences between the estimates of baseline length and their weighted mean from 14 experiments with the Haystack-Owens Valley interferometer. The weighted mean of these estimates of baseline length was 392,888,164.4 cm. The value of the speed of light used to convert light travel time to centimeters was $2.99792458 \times 10^{10}$ cm/s. The source coordinates were also estimated in these analyses.

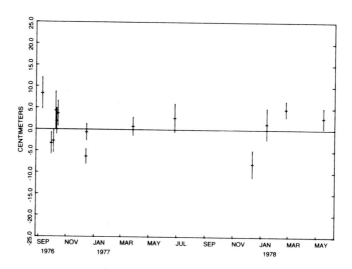

Figure 2. Same as Figure 1, but with each source coordinate kept fixed at the same value for all analyses (see text). The weighted mean of these length estimates was 392,888,162.6 cm.

scatter to be reduced, provided that any systematic errors affecting the data and the model are sufficiently benign. The estimates of baseline length from these solutions are shown in Figure 2; the RMS scatter is indeed reduced, to about 4 cm, and the mean itself, as one might expect, changed little, by only 1.8 cm. These results indicate that at present VLBI can be used to determine lengths of transcontinental baselines with a repeatability of about 4 cm.

It is not possible from VLBI data alone to determine the point at which the rotation pole of the Earth pierces its crust. It is possible, however, to measure changes in this position from the position at some arbitrary epoch, the position of the pole at that epoch being used to define the orientation of the baseline in the Earth-fixed frame. The ability to determine these changes is, therefore, related to the ability to determine the orientation of the baseline. An examination of the relevant error ellipsoids indicates that, for our interferometer, the inherent sensitivity of the VLBI data to baseline orientation is about a factor of four less than their sensitivity to baseline length. We would expect, therefore, that uncertainties in the determinations of changes in the position of the pole from these data would be no greater than about 30 cm, equivalent to about 0".01 uncertainty in the determination of the direction of the pole.

To determine changes in the position of the pole, we combined all of the data in a single analysis and estimated these changes and all other relevant parameters simultaneously. For one experiment, conducted on 4 October 1976, the position of the pole was fixed at the value determined by the BIH, as recorded in their Circular D publications, in order to define the orientation of the baseline. Our results for the position of the pole from the other 13 experiments are shown in Figure 3 in the form of the difference between the VLBI and the BIH values.

Also plotted in Figure 3 are the differences from the BIH values of the x component of the pole position determined by the IPMS from optical observations (tabulated at intervals of 0.05 years), and by the Polar Monitoring Service of the Defense Mapping Agency Topographic Center from satellite Doppler observations (tabulated at intervals of 5 days), both as recorded in the Series 7 publications of the U. S. Naval Observatory. The standard deviations for the IPMS data are typically about 50 cm or 0".015. The changes with time for all three sets have a common trend, systematically different from the BIH values. This common trend appears to have persisted at least through May 1978. Thus, the VLBI results seem to agree with both the IPMS results and the Doppler results better than any one of the three sets agrees with the BIH values. The RMS scatter of the differences between the VLBI and BIH determinations shown in Figure 3 is 0".030; the corresponding scatter for the Doppler results is 0".020; and for the IPMS results 0".027. This scatter in the VLBI results is about 30 percent less than that obtained previously (Shapiro et al., 1974; Robertson, 1975) from the analysis of VLBI observations made in 1972 and 1973.

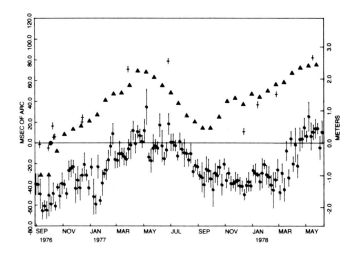

Figure 3. x-component of the pole position, from VLBI observations (+) of extragalactic radio sources, from Doppler observations (¢) of satellites, and from IPMS optical observations (▲) of Galactic stars, all relative to the values obtained by the BIH (see text). In the analysis of the VLBI data, the pole position was fixed at the BIH value for the experiment of 4 October 1976 (●).

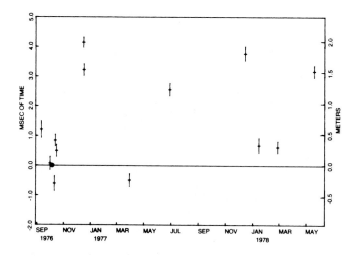

Figure 4. Comparison of the VLBI and BIH determinations of UT1. The VLBI value for UT1 for the experiment of 4 October 1976 was fixed at the BIH value (●).

Figure 4 shows the corresponding comparison of the VLBI and BIH results for UT1, obtained from the same analysis, with UT1 having been set at the BIH value for 4 October 1976. The RMS scatter here, about $0\overset{s}{.}002$, is again less than that obtained previously (Shapiro et al., 1974; Robertson, 1975) and is about the same as the scatter seen in the UT1 determinations from lunar laser ranging (King et al., 1978; see also Stolz et al., 1976).

In this paper we have summarized our current ability to determine baseline lengths, pole position, and Earth rotation with VLBI. Another paper presented at this conference (Carter et al., 1979) addresses plans we have for the improvement of both the quality and the quantity of such determinations.

REFERENCES

Carter, W. E., Robertson, D. S., and Abell, M. D.: 1979, this volume.
Clark, T. A., Hutton, L. K., Marandino, G. E., Counselman III, C. C., Robertson, D. S., Shapiro, I. I., Wittels, J. J., Hinteregger, H. F., Knight, C. A., Rogers, A. E. E., Whitney, A. R., Niell, A. E., Ronnang, B. O., and Rydbeck, O. E. H.: 1976, Astron. J. 81, pp. 599-603.
Guinot, B.: 1970, Astron. Astrophys. 8, pp. 26-28.
Guinot, B.: 1974, Astron. Astrophys. 36, pp. 1-4.
King, R. W., Counselman III, C. C., and Shapiro, I. I.: 1978, J. Geophys. Res. 83, pp. 3377-3381.
Lieske, J. H., Lederle, T., Fricke, W., and Morando, B.: 1977, Astron. Astrophys. 58, pp. 1-16.
Melchior, P.: 1971, Celes. Mech. 4, pp. 190-212.
Robertson, D. S.: 1975, PhD. Thesis, M.I.T.
Shapiro, I. I., Robertson, D. S., Knight, C. A., Counselman, C. C., Rogers, A. E. E., Hinteregger, H. F., Lippincott, S., Whitney, A. R., Clark, T. A., Niell, A. E., Spitzmesser, D. J.: 1974, Science 186, pp. 920-922.
Stolz, A., Bender, P. L., Faller, J. E., Silverberg, E. C., Mulholland, J. D., Shelus, P. J., Williams, J. G., Carter, W. E., Currie, D. G., and Kaula, W. M.: 1976, Science 193, pp. 997-999.
Woolard, E. W.: 1953, Astron. Papers of the Amer. Eph. 15, part 1.

DISCUSSION

J. D. Mulholland: The equations given by Johnston suggest that your claim to solve directly for UT1 assumes that the Z-component of the baseline is known without error.
W. E. Carter: The determination of UT1 is relatively insensitive to small errors in Z.

A. R. Robbins: I believe that tapes have to be played back in real time for cross-correlation, and that very few organizations can carry this out. Consequently it takes a very long time to get results from experiments. Can you comment on this?

W. E. Carter: The Mark III VLBI system, which will be utilized in project POLARIS, has been designed with a high degree of automation of the data handling and processing which will allow the solutions to be completed in a matter of days after receipt at the processing facility. In the future, communications satellites may be utilized, and the results could be determined nearly in real time.

K. Johnston: What form of atmospheric correction did you apply to the data? If H_2O radiometers had been used to measure the H_2O atmospheric content along the line of sight, how would this have improved the accuracy of your results?

W. E. Carter: Only surface meteorological measurements were used with a model derived by J. Marini at NASA. It appears that H_2O radiometers may reduce the uncertainty in the atmospheric corrections to the sub-centimeter level.

PRECISION ESTIMATES OF UNIVERSAL TIME FROM RADIO-INTERFEROMETRIC
OBSERVATIONS

H. G. Walter

Astronomisches Rechen-Institut, Heidelberg, F.R.Germany

ABSTRACT

Considerable improvement in the determination of the motion of the Earth is possible by the potentially high accuracy inherent in very-long-baseline interferometry. Precisions of UT1 are estimated from time delay and fringe frequency measurements of extragalactic radio sources with positional uncertainties at the $0\overset{''}{.}01$ level. Case studies resulted in standard deviations about one order of magnitude smaller than those obtained by classical astrometric methods. The dependence of estimates on baseline orientations and source declinations is discussed.

1. INTRODUCTION

Standard deviations of UT1 as deduced by BIH from classical astrometric observations amount to $\sigma(UT1) = 0\overset{s}{.}001$ for a 5 day mean average (Guinot, 1970; Feissel et al., 1972). Lunar laser ranging and very-long-baseline interferometry (VLBI) are promising techniques with the potential to further improve the accuracy. Currently, these techniques have reached a state which is competitive with classical methods: UT1 determination from lunar laser ranging, VLBI and BIH data indicates agreement at the level of 1 to 2 ms (King et al., 1977). Prospective improvements of VLBI techniques will enable measurements of UT1 with uncertainties equivalent to a few centimeters and with time-resolution finer than one day as compared with five day averages hitherto (Counselman, 1976).

An attempt is made to substantiate the predicted accuracy for time from VLBI measurements by a least squares estimate. In pursuing the approach realistically, baselines formed by the antennae of the Deep Space Network are mainly employed as they also figure in NASA's project for the determination of UT1, polar motion and clock synchronisation. Within the framework of this paper the dependence of parameter estimation on the declination of observed radio sources, the choice of observables, i.e. time delay and/or fringe frequency, and the geographical location of baseline terminals is investigated on the assumption of known positions of radio sources which constitute an adopted radio reference frame.

2. SIMULATION AND PROCESSING OF OBSERVATIONS

In contrast to a rigorous treatment of real radio interferometric observations requiring corrections accounting for environmental and instrumental effects it is practical to restrict a parameter estimation study to a simplified model, which is solely defined by the interferometer geometry. For this purpose an inertial coordinate system with origin at the centre of the Earth is constructed. Vectors to the antennae are designated $\vec{r}_i(t)$, $i = 1, 2, \ldots$. Let $\vec{r}_s$ be the unit vector to the radio source. Then, under these purely geometrical conditions, the zero-order baseline $\vec{b}$, time delay τ and fringe frequency f are defined by

$$\vec{b} = \vec{b}(t) = (\vec{r}_i - \vec{r}_k) = d\,\vec{r}_b, \qquad (1)$$

where d denotes the baseline length and $\vec{r}_b$ the unit vector along the line interconnecting two antennae;

$$\tau = \frac{d}{c}\,\vec{r}_s\,\vec{r}_b \qquad (2)$$

and

$$f = \omega\,\frac{d}{c}\,\vec{r}_s\,\dot{\vec{r}}_b, \qquad (3)$$

which holds for radio sources of negligible proper motion such as extragalactic sources. Here, ω denotes the signal frequency and is kept constant at the level of 2.3 GHz. For details on radio interferometry and astrometrical aspects reference is made to Cohen and Shaffer (1971) and Counselman (1976).

Observations were simulated for samples consisting of 8 radio sources and 2 or 3 baselines. In order to discover possible dependences of parameter estimations on source declinations the sources were subdivided in three groups, G1, G2, G3 ranging from $0° < \delta < +25°$, $+20° < \delta < +45°$ and $-20° < \delta < +5°$, respectively. G1 comprises the sources 3C 120, 3C 138, 0736+01, OJ 287, 3C 273, OQ 208, CTA 102, 3C 454.3; G2 the sources 3C 48, 3C 84, OJ 287, 4C 39.25, 3C 286, OQ 208, 3C 345, BL Lac; G3 the sources CTA 26, 3C 120, 0736+01, 3C 273, 3C 279B, 1741-038, OX 057, 2345-16.

Likewise, a variety of baseline configurations was chosen to examine the influence of large equatorial versus large polar baseline components on the variances and covariances. Essentially, the study rests on baselines constituted by the antennae of NASA's Deep Space Network (Goldstone, Calif.; Robledo, Spain; Tidbinbilla, Australia), tentatively augmented by the radio telescope located at Hartebeesthoek (South Africa).

Whenever mutual visibility for a source from the baseline terminals was secured observation times in steps of about 30 minutes were generated throughout the visibility periods during 24 hours. The observations were assumed normally distributed with mean zero and uncorrelated with standard deviations of $\sigma(\tau) = 10^{-9}$ s and $\sigma(f) = 10^{-4}$ Hz, which corresponds to a source position uncertainty of the order of $0\overset{''}{.}01$ for a baseline length of 5000 km.

The estimation of UT1 necessitates the setting up of the partial derivatives of the observables with respect to Greenwich sidereal time α_G. As the instantaneous baseline vector $\vec{b}$ is referred to the geographical position vectors $\vec{\rho}_1$ and $\vec{\rho}_2$ of the terminals through

$$\vec{b} = R_G R_P (\vec{\rho}_2 - \vec{\rho}_1)_{CIO} \qquad (4)$$

the required partial derivatives are obtained from Eqs. (1) and (2) by differentiating $\vec{b}$ and $\dot{\vec{b}}$ accordingly. The matrix R_P accounts for polar motion, whereas the matrix R_G rotates the baseline vector to its instantaneous orientation by the angle $\alpha_G = \alpha_G(t)$, the right ascension of Greenwich.

3. DISCUSSION OF RESULTS

Variances averaged over one day are deduced from 100 observations each for different sets of radio sources and baselines from customary covariance matrix calculations (e.g. Liebelt, 1967).

For sources subjected to the criterion of simultaneous visibility from terminals the baseline Tidbinbilla-Robledo did not yield noteworthy source availability times and was, therefore, omitted. Thus, the parameter estimates concentrated on the following baseline configurations: Robledo-Goldstone, Goldstone-Tidbinbilla, (I), both of which are baselines with predominant equatorial projections; Robledo-Goldstone, Goldstone-Tidbinbilla, Hartebeesthoek-Robledo, (II), where the last baseline is characterized by a large polar component; Robledo-Goldstone, Hartebeesthoek-Robledo, (III), in which configuration the baselines of sizable equatorial and polar components have an equal share; Goldstone-Tidbinbilla, Hartebeesthoek-Robledo, (IV), with baselines of large polar-axis projection.

The standard deviations of UT1 for a three parameter model are depicted in Figure 1 arranged by source declinations and baseline configurations. While the standard deviations of UT1 for a combined solution of an equal number of time-delay and fringe-frequency observations are practically independent of source declinations they exhibit noticeable differences concerning the baseline configuration. Baselines with large polar-axis projection cause unfavourable standard deviations as demonstrated by configuration IV, whereas baselines with large equatorial projections provide the better estimates.

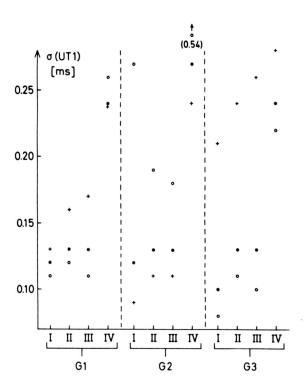

Fig. 1. Dependence of UT1 estimates on source declination (G1 to G4), baseline network (I to IV) and observables (τ : o, f : +, τ & f : ●)

This fact is even more pronounced when only one type of observable, τ or f, is analysed. Evidently, the determination of UT1 from time delay observations of sources in the equatorial zone (e.g. G3) is more precise than from any other distribution of source declinations. The results obtained for fringe frequency, however, suggest that UT1 is more sensitive when sources of higher declinations such as G2 participate in the observation campaign.

Since a covariance analysis is closely related to the derivatives of the observables with respect to the parameters to be estimated a partial explanation of the behaviour of the standard deviations, at least, can be expected from the analysis of the derivatives. Williams (1970) has performed such analysis for fringe frequency observations. Accordingly, UT1 is most accurately determined for sources at zero-degree declinations and improves with increasing equatorial projection of the baseline.

These results are confirmed by the current covariance analyses with the exception that the superiority of UT1 estimates, although minimal, from fringe frequency observations of sources in the zone $+20° < \delta < 45°$ contradicts the conclusions from partial derivatives which predict detection of the best estimates from sources at zero-degree declinations (e.g. G1 or G3). Presumably the discrepancy can be attributed to the choice of source declinations from a narrow zone around zero-degree, thus giving rise to near - linearity of observational partial derivatives and leading to ill-conditioned covariance matrices.

4. CONCLUSION

Table 1 summarizes the present state of accuracy for UT1 and lists the prospective estimates deduced from observations obtained through VLBI. In the course of the estimation process no allowance was made for signal disturbance caused by the propagation-medium, the instrumentation

Method of Observation	$\sigma(UT1)$ [s]
Classical astrometry 5 day resolution (Guinot, 1970; Feissel et al., 1972)	0.001
Lunar Laser Ranging 5 day resolution (Harris and Williams, 1977)	0.0014
Lunar Laser Ranging (King et al., 1977)	~ 0.001
VLBI - pilot study (Shapiro et al., 1974)	0.002
VLBI - parameter estimate 1 day resolution (Moran, 1973)	0.0008
VLBI - parameter estimate 1 day resolution (c.f. this paper)	0.0002

Table 1. Standard errors of UT1

and the clock offset. Further, it was assumed that the effects of Earth tides, continental drift and possible others were properly accounted for in the data reduction. For these reasons the estimates may have turned out too optimistic and are likely to deteriorate under real conditions. To the extent that the current results are representative, determination of UT1 by VLBI techniques appears to be most effective for time-delay and fringe-frequency observations of sources in the equatorial zone acquired from a net of baselines in which large equatorial and polar-axis projections are kept in balance.

REFERENCES

Cohen, M.H., Shaffer, D.B.: 1971, Astron. J. 76, pp. 91-100.
Counselman III, C.C.: 1976, Annual Rev. of Astron. and Astrophys. 14, pp. 197-214.
Guinot, B.: 1970, in "Earthquake Displacement Fields and the Rotation of the Earth", ed. L. Mansinha, D.E. Smylie, A.E. Beck; D. Reidel Publ. Comp., Dordrecht, Holland, pp. 54-62.
Harris, A.W., Williams, J.G.: 1977, in "Scientific Applications of Lunar Laser Ranging", ed. J.D. Mulholland; D. Reidel Publ. Comp., Dordrecht, Holland, pp. 179-190.
King, R.W., Clark, T.A., Knight, C.A., Counselman III, C.C., Robertson, D.S., Shapiro, I.I.: 1977, in "Scientific Applications of Lunar Laser Ranging", ed. J.D. Mulholland; D. Reidel Publ. Comp. Dordrecht, Holland, pp. 219-220.
Liebelt, P.B.: 1967, An Introduction to Optimal Estimation, Addison-Wesley Publ. Comp., Reading, Massachusetts.
Moran, J.M.: 1973, in "Space Research XIII", Vol. 1, ed. M.J. Rycroft, S.K. Runcorn; Akademie-Verlag, Berlin, pp. 73-82.
Shapiro, I.I., Robertson, D.S., Knight, C.A., Counselman III, C.C., Rogers, A.E.E., Hinteregger, H.F., Lippincott, S., Whitney, A.R., Clark, T.A., Niell, A.E., Spitzmesser, D.J.: 1974, Science 186, pp. 920-922.
Williams, J.G.: 1970, Space Programs Summary 37-62, Vol. II, Jet Propulsion Laboratory, Pasadena, Calif., pp. 37-50.

DISCUSSION

K. Johnston: Did your calculations include limitations due to source structure?

Dr Carter presented data in which the accuracy of the baseline was about 5 cm; this is better than the value you have assumed in your calculations.

H.G. Walter: No limitations due to source structure were included, but the assumed precision of the source positions was set at 0.01 seconds of arc.

A time delay precision of 1 ns has been tentatively assumed for this computer simulation; it could be improved to the level quoted by Dr Carter.

PART V : SATELLITE LASER RANGING

DETERMINATION OF POLAR MOTION AND EARTH ROTATION FROM LASER TRACKING OF SATELLITES

David E. Smith and Ronald Kolenkiewicz
Geodynamics Branch, Goddard Space Flight Center
Greenbelt, Maryland
Peter J. Dunn and Mark Torrence
EG&G, Washington Analytical Services Center, Inc.
Riverdale, Maryland

ABSTRACT

Laser tracking of the Lageos spacecraft has been used to derive the position of the Earth's pole of rotation at 5-day intervals during October, November and December 1976. The estimated precision of the results is 0.01 to 0.02 arcseconds in both x and y components, although the formal uncertainty is an order of magnitude better, and there is general agreement with the Bureau International de l'Heure smoothed pole path to about 0.02 arcseconds. Present orbit determination capability of Lageos is limited to about 25 cm rms fit to data over periods of 5 days and about 50 cm over 50 days. The present major sources of error in the perturbations of Lageos are Earth and ocean tides followed by the Earth's gravity field, and solar and Earth reflected radiation pressure. Ultimate accuracy for polar motion and Earth rotation from Lageos after improved modeling of the perturbing forces appears to be of order ± 5 cm for polar motion over a period of about 1 day and about ± 0.2 to ± 0.3 milliseconds in U.T. for periods up to 2 or 3 months.

INTRODUCTION

The determination of polar motion and Earth rotation from the tracking of satellites is based on the concept that over a given period of time the motion of the satellite about the Earth is known with sufficient accuracy that for all practical purposes it is fixed in an inertial (reference) frame. In reality, the spacecraft is continuously perturbed by forces interior and exterior to the Earth and the orbit of the spacecraft is continuously evolving. The use of satellite orbits for polar motion and Earth rotation, therefore, reduces primarily to a problem in the determination of the spacecraft orbit and to the detailed understanding of the changes that the orbit goes through. Thus, key aspects are the process of orbit determination from tracking data and orbital stability.

The problems of orbit determination are associated with the quality, quantity, and distribution of the tracking. Frequently, some or all of these factors are outside the control of the scientist using the data, but it is possible to plan and specify the requirements in each of these areas if the final quality of the required orbit is known. The orbit determination process is essentially deterministic and generally well understood although opinions may differ on the optimum mathematical techniques that should be used. The limiting factor in satellite techniques for polar motion and Earth rotation is clearly our degree of understanding (or predictability) of the evolution of the orbit over a period of time. Fortunately, the primary forces perturbing the orbits of satellites are known, and consequently it has been possible to design a spacecraft and its orbit to minimize the influence of these forces. This spacecraft, Lageos (Laser Geodynamics Satellite), was launched on May 4, 1976 into a high orbit nearly 6000 kilometers above the Earth's surface, and because of its altitude, is much less affected by poorly-known short wavelength features in the gravity field. The spacecraft is heavy (411 kg) and therefore almost unaffected by the perturbing forces of solar radiation, Earth albedo and air drag which severely perturb spacecraft of less weight at lower altitudes. Thus the Lageos spacecraft and orbit come close to acting as an artificial reference system in Earth orbit. A summary of Lageos and its orbit is given in Table 1.

Table 1. Lageos Spacecraft and orbit.

Launch:	May 4, 1976	
Spacecraft:	Spherical, 60 cm diameter	
	411 kg	
	426 retro-reflectors, each 3.8 cm in diameter	
Orbit:	Semi-major axis	12265 km
	Inclination	109.8 degrees
	Eccentricity	0.004
	Perigee height	5858 km
	Apogee height	5958 km

Initial experiments to determine polar motion from laser tracking of satellites were conducted using the Beacon Explorer C spacecraft at 1000 km altitude. We were able to show that the variation of latitude could be determined at about the 1 meter level with this spacecraft over periods as long as eighteen months and that an estimate of Earth rotation could be derived over a short period of 3 weeks (Smith, et al., 1972; Kolenkiewicz, et al., 1974; Dunn, et al., 1977). It was clear, however, from these studies that perturbations of the spacecraft by the Earth's gravity field, Earth and ocean tides, air drag, etc., were limiting the capability and that a higher and heavier spacecraft (Lageos) was necessary. Further, all these investigations had been conducted with a single laser station and this was an additional limitation on the development of the technique.

The method of deriving polar motion from a satellite laser tracking network is different from the single station case because with a network, long-term stability of the orbit is not a requirement. Orbit stability is only required over the period (hours to days) over which a single pole position is desired. That is, for a five-day (mean) pole position orbit stability is only required over the five days. With a network of stations the determination of the pole position reduces to deriving the coordinates of that point (the pole) about which the network appeared to rotate during the period (five days, one day, etc.). From one period to the next there is no requirement for the orbit to be known; indeed, a different satellite can be used since the satellite is only a common object for all the tracking stations to observe. Common to all pole position determinations is the station coordinates of the laser systems and these must be well-known. It is in the coordinate system of the stations that the pole position is referred. For Earth rotation measurements (time), however, orbit stability is essential. The means by which Earth rotation can be derived from a satellite is to measure the relative rotation about the Earth's spin axis of the network (or station) with respect to the orbital plane of the spacecraft. In order for this measurement to be referred to an inertial frame the orbital plane must be unperturbed or its perturbations very predictable. Although the perturbations of the Lageos orbit are small the slow secular perturbation of the node of the orbit by effects not fully understood will always exist. Thus short-term variations in Earth rotation will be more easily (and accurately) observed with Lageos than long term changes. Further discussion of the measurement of polar motion and Earth rotation from spacecraft is given in Kolenkiewicz, et al. (1977).

LAGEOS RESULTS

The first eight months of tracking of the Lageos spacecraft by the Goddard Space Flight Center (GSFC) and Smithsonian Astrophysical Observatory (SAO) laser systems have been analyzed. During this period (May to December 1976) the GSFC lasers operated from sites in North America: Greenbelt, MD (Stalas); San Diego, CA (Moblas 3); Quincy, CA (Moblas 2); Bear Lake, UT (Moblas 1). During the first few months only the Stalas system was operational and during the last three months only the three lasers at San Diego, Quincy, and Bear Lake were in operation. The SAO lasers were located at Arequipa, Peru; Natal, Brazil; Orroral, Australia; and Mt. Hopkins, AZ in North America. The Orroral system did not begin operations until the latter part of 1976.

The quality of the data obtained from these systems was approximately 10 cm rms deviation for a single measurement from the GSFC system and approximately 1 meter from the SAO systems. The quality of orbit determination with the tracking data is summarized in Table 2. With the better data orbital fits of a few tens-of-centimeters were obtainable for orbital arcs of one month or more. However, it should be remembered that the three GSFC lasers that were used in the orbit determination were all located within about 1200 km of each other, and we therefore cannot be sure that similar orbital fits would be obtainable if one of these stations had been in Australia. An upper limit on the "poorness" of these orbits is provided by the SAO Orroral laser in which orbital fits of better than 1 meter were regularly obtained with orbital arcs of 5 days in length.

Table 2. Lageos orbit fits with GSFC laser data.

	Orbital Fit
1 pass (45 minutes)	10 cm
5 days	25 cm
30 days	40 cm
50 days	50 cm

From the tracking data a set of 31 five-day orbital arcs were determined from which a set of station coordinates was derived (to be published elsewhere). Using these station coordinates 30-day orbital arcs were derived for each of the months, October, November and December 1976. During each 5-day period within each monthly arc the station network was rotated about the equatorial axes through Greenwich and 90° W longitude (parallel to the x and y polar coordinate axes) in order to better fit the tracking data to the orbit and thus provide an improved mean position of the pole during the 5-day period. The 30-day orbit of the spacecraft was then re-adjusted using the up-dated pole positions and each of the pole positions re-determined. This iterative procedure was continued until no change in the orbit or the pole was detectable between iterations. We found that the solution converged to better than 10^{-3} arcseconds after only one iteration.

The x and y polar coordinates obtained from the Lageos data in the manner described are shown in Figures 1 and 2. Near the end of December there were only a few observed passes of the satellite. The recovered y value of the pole position for December 27 was very different from previous values and has been omitted from Figure 1. The formal uncertainty of each of the x values is approximately 3×10^{-3} arcseconds and approximately 2×10^{-3} arcseconds for y. Some variation in the uncertainty exists between the points and reflects the quantity and quality of the data. The y component is more strongly determined because most of the stations had longitudes in the region of 90° W.

The true uncertainty in both x and y is believed to be between 1 and 2×10^{-2} arcseconds for most of the data.

Figures 1 and 2 show good general agreement between the smoothed BIH values and the Lageos results for most of the three-month period. However, we do not believe the large departure in x from the smoothed

BIH in December 1976 is real. At the present time we do not have an explanation of this difference but stress that the Lageos results presented here are preliminary. The results described were obtained using the GEM 10 gravity field model (Lerch, et al., 1977) and the Geodyn orbit determination program.

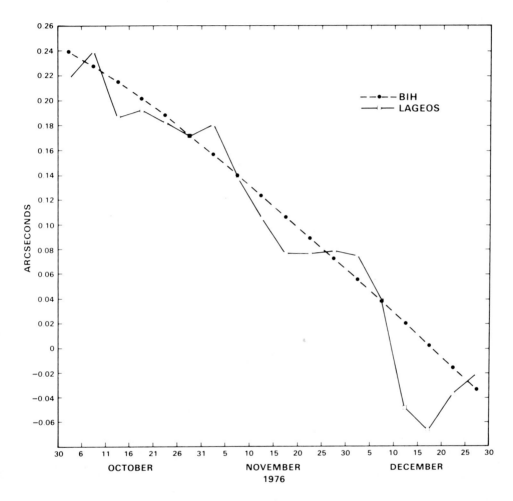

Figure 1. X-component of polar motion.

LIMITATIONS

As described in the Introduction the accuracy of polar motion and universal time determined from Lageos tracking data is limited by orbit stability. Simulations that we have performed (Kolenkiewicz, et al., 1977) indicate accuracies of the order of 1 or 2 x 10^{-3} arcseconds in polar motion are possible over a few days even from a single station

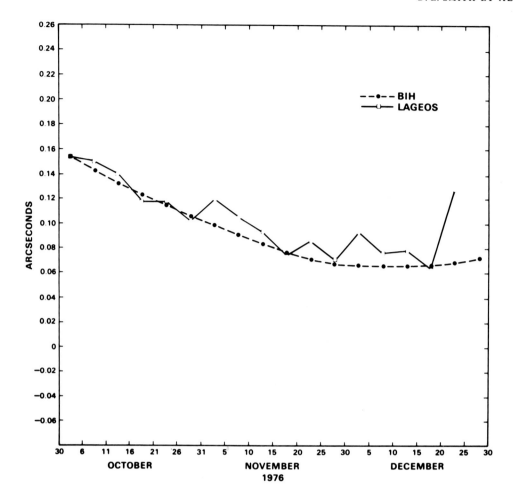

Figure 2. Y-component of polar motion.

and can find no perturbing forces which, in principle, cannot be adequately modeled for Lageos at the 10^{-3} arcsecond level. These include gravity, station coordinates, Earth mass, solar radiation pressure, and Earth albedo radiation, although the latter is extremely complex and difficult to assess (Smith, 1970). For Earth rotation the limitation appears to be albedo and ocean tides. Our estimate of the nodal perturbation by uncertainties in albedo is a few thousandths of a second of arc after a few months based on a comparison with uncertainties in direct solar radiation effects which we always find to be greater than, or comparable to, albedo. Ocean tides are at present even less well known but will probably be completely modelable after a few years of tracking. The ocean (and solid Earth) tides perturb the orbital inclination and node, producing periodic and near-secular terms in the latter. These terms in the acceleration of the orbit may be the present limitation on using Lageos for measuring U.T. Our present assessments of these forces on the orbital inclination and node are shown in Table 3.

Table 3. Lageos orbit perturbations during first 230 days after launch.

	Error	Orbital Inclination (degrees)	Nodal Position (degrees)
Solar radiation pressure	10%	3×10^{-7} (3cm)	4×10^{-7} (4cm)
Earth albedo pressure	100%	4×10^{-7} (4cm)	6×10^{-7} (7cm)
Earth and ocean tides	10%	6×10^{-5} (7m)	2×10^{-5} (2m)
Gravity	(*)	3×10^{-7} (3cm)	1.5×10^{-6} (17cm)

*Difference between Goddard Models 9 and 10 (GEMs 9, 10)

The inclination is probably the single most important parameter for estimating polar motion, and the node for estimating U.T. It should be noted that Table 3 shows the perturbation during a 230-day period, and that for polar motion periods of 5 days or less, the perturbation will be considerably smaller for all secular or long period effects. If the ocean tides can be adequately modeled then the ultimate polar motion accuracy is probably of the order 5 cm, over averaging times from 1 to 5 days, and about 0.2 to 0.3 milliseconds in U.T. over 3 months. The date when this capability will be approached will probably be around 1981-2, assuming continued development of systems and models.

CONCLUSIONS

Five-day polar motion has been derived from Lageos laser tracking for the period October through December 1976 with a precision of 0.01 to 0.02 arcseconds. Although the data were frequently of only 1 meter quality, never more than seven stations tracking, and the period only 3 months duration, these preliminary results demonstrate that Lageos can be used for determining polar motion as originally envisaged prior to the launch of the spacecraft. Improvements in the quality of the results can be expected over the next few years as more stations begin to track Lageos at the 10 cm level, as our modeling of the perturbing forces improves (particularly Earth and ocean tides), and as the quantity and regularity of the tracking increases. As far as can be determined at the present time there appears to be no perturbing forces which should not be adequately modelable in the next few years to ultimately permit 5 cm daily polar motion determinations and 0.2 millisecond U.T. measurements.

REFERENCES

Dunn, P. J., Smith, D. E., and Kolenkiewicz, R.: 1977, J. Geophys. Res. 82, pp. 895-897.
Kolenkiewicz, R., Smith, D. E., and Dunn, P. J.: 1974, in G. Veis (ed.), "The Use of Artificial Satellites for Geodesy and Geodynamics", National Technical University of Athens, Athens, pp. 813-823.
Kolenkiewicz, R., Smith, D. E., Rubincam, D. P., Dunn, P. J., and Torrence, M. H.: 1977, Phil. Trans. R. Soc. Lond. A. 284, pp. 485-494.
Lerch, F. J., Klosko, S. M., Laubscher, R. D., and Wagner, C. A.: 1977, "Gravity Model Improvement Using GEOS-3 (GEM 9 and 10)", Goddard Space Flight Center, X-921-77-246.
Smith, D. E.: 1970, in B. Morando (ed.), "Dynamics of Satellites", Springer-Verlag, Berlin, pp. 284-294.
Smith, D. E., Kolenkiewicz, R., Dunn, P. J., Plotkin, H. H., and Johnson, T. S.: 1972, Science 178, pp. 405-406.

DISCUSSION

P. Paquet: What is the present rms error of the 5-day polar coordinates obtained by this method?

D. E. Smith: The internal precision of each 5-day value is approximately 0.001 arcsec. The consistency between the 5-day values is about $0\overset{''}{.}01$ rms and the rms deviation with respect to BIH values is nearly $0\overset{''}{.}02$.

P. Paquet: You quoted errors of between 20 cm and 50 cm in the station coordinates. How many days of observations were used in these determinations?

D. E. Smith: 150 days of observations were used to determine the coordinates of all the tracking stations. Three sub-sets, each of 50 days, showed differences ranging from a few centimeters to over 50 cm. The poorest determinations were for some of the Smithsonian stations that only had one meter ranging capability. I estimate the general accuracy of this network to be near 20 cm for the better stations and about 50 cm for the weaker stations.

P. Brosche: Which model of oceanic tides was used? I recommend Dr. Zahel's newest model including self attraction and loading.

D. E. Smith: We did not model the ocean tides - only those of the solid Earth. However, it is clear from our investigations that the ocean tides must be included and we shall try several models on the Lageos data.

P. Brosche: What is the reason that the satellite laser techniques does not give tremendously better results than lunar laser ranging?

E. C. Silverberg: This is a common misunderstanding brought about by the fact that the satellite workers report "accuracy" as their single-shot uncertainty, while lunar workers average many shots and report normal point "accuracy". In fact, the single-shot uncertainty is usually numerically greater for lunar systems than for satellite systems, but the slowness of the lunar orbit permits more averaging of data

POLAR MOTION FROM LASER RANGE MEASUREMENTS OF GEOS-3

B. E. Schutz, B. D. Tapley, J. Ries
Department of Aerospace Engineering and Engineering Mechanics
The University of Texas at Austin
Austin, Texas

ABSTRACT

Using two-day arcs of GEOS-3 laser data, simultaneous solutions for pole position components, x_p and y_p, and orbit elements have been obtained for the period spanning 3 February to 6 March 1976 using three NASA Goddard Space Flight Center laser stations located near Washington, D.C. (STALAS) and on the islands of Bermuda and Grand Turk. The results are in general agreement with the BIH results. However, because of the locations of the laser sites, the x_p solution is weaker than the y_p solution. The x_p and y_p estimates were smoothed with a straight line by weighted least squares using the variance associated with the pole estimates as weights in order to reflect the effect of widely different data distributions. The smoothed y_p differs by one meter with respect to the BIH smoothed values and the smoothed x_p differs by about two meters. Spectral analysis of the results has identified frequencies associated with the orbital motion indicating the need for further improvements in the model of the physical system.

INTRODUCTION

The launch of the Geodynamics Experimental Ocean Satellite (GEOS-3) on April 9, 1975, initiated a priority laser tracking campaign to provide the necessary data for determining an accurate orbit of the satellite. These lasers included the NASA Goddard Space Flight Center (GSFC) systems which operated at a precision of 5 to 10 cm, the Smithsonian Astrophysical Observatory (SAO) network of stations, and participating systems in other countries. Although the motivation for the laser deployment was to provide an accurate orbit for calibration and use of the GEOS-3 altimeter, it was apparent that contributions could be made to other applications requiring high precision such as polar motion. Previous polar motion experiments with laser systems were performed by Smith, et al. (1972) and Dunn, et al. (1977) using the Beacon Explorer-C Satellite to obtain one component of polar motion. Thus GEOS-3 offered an opportunity to obtain polar motion results using intensive tracking from several high precision laser ranging systems.

A single laser range measurement may be described mathematically as a function of the geocentric coordinates of the satellite center of mass, the coordinates of the electronic center of the laser system, and a term representing contributions from the atmospheric refraction, noise and bias in the laser system, the correction between the reflector location and the center of mass of the satellite, errors resulting from an incomplete model of the dynamical system and the station coordinates errors. The commonly chosen origin for the reference geocentric coordinate system is the Conventional International Origin (CIO) as defined by the coordinates of the International Latitude Service stations. If the CIO system is used in conjunction with laser range measurements, an inherent problem exists to determine the laser coordinates in the CIO system.

Polar motion influences range measurements in two ways, viz., dynamically and kinematically. The former polar motion effect produces a force on the satellite as the result of changes in the inertial orientation of the geopotential. The latter effect is dominant and is dependent on the angle between the angular velocity vector of the Earth and the geocentric position vector of the observing station. As shown by Lambeck (1971), the kinematic influence on the laser range measurement includes a nearly diurnal signature. Consequently, a polar motion determination from artificial satellites requires proper modeling of all diurnal or nearly diurnal perturbations as well as other parameters which may appear to be diurnal as the result of non-uniform sampling of the orbit by the laser.

GEOS-3 MODEL AND DATA

The GEOS-3 laser range data have been analyzed using the University of Texas Orbit Processor (Wilson, 1978) using a force model which included the GEM-10 geopotential (Lerch, et al., 1977), the luni-solar perturbations using DE-96 (Standish, et al., 1976), solid Earth tides with $k_2 = 0.306$ (Farrell, 1972), atmospheric drag using Modified Harris-Priester (Dowd, 1977), and solar radiation pressure. The coefficient of drag and its time derivative were estimated in one set of solutions while C_D was held fixed to 2.3 in a comparative set.

The laser measurement model used in the analysis included laser station coordinates from GEM-10, station tides with $h_2 = 0.615$ and $\ell_2 = 0.084$ (Gutenberg-Bullen A Earth model, Farrell, 1972), tropospheric refraction corrections, laser reflector center-of-mass corrections, and appropriate timing corrections. Although a separate solution was obtained for station coordinates using various geopotentials, the GEM-10 set was adopted to maintain consistency with the geopotential, particularly with respect to terms producing signatures which are similar to polar motion.

The laser data used in the analysis were obtained from three NASA Goddard Space Flight Center stations located at GSFC (STALAS) and on the islands of Bermuda and Grand Turk. The data analyzed were obtained during an intensive GEOS-3 tracking campaign from 3 February 1976 (760203) to 6 March 1976 (760306). This 32-day time period was divided into two-day arcs in which solutions for the pole components, x_p and y_p (assumed constant over the arc), and the orbital elements at the beginning of the arc were estimated. Two-day arcs were selected to ensure an adequate distribution of laser data in time while simultaneously reducing the effect of unmodeled orbital perturbations. Each arc was chosen to begin at $0^h0^m0^s$ UTC(BIH) on the appropriate date. For the sixteen arcs considered, laser measurements from all three stations were obtained within each arc except for the arcs beginning 760203, 760205, 760207, 760302 and 760304, in which no measurements from STALAS were obtained. On eleven occasions during this 32-day period, all three stations observed the same orbital revolutions. In addition, the measurement precision was less than ten centimeters and the systems pulsed once per second; however, an average pass produced between 150 and 200 measurements from a single station. Finally, although some GEOS-3 laser data were included in the GEM-10 geopotential and station coordinate solution, the February 1976 data were not used.

RESULTS

The estimates for the polar motion component y_p are shown in Figure 1, and Figure 2 gives the root mean square (rms) of the laser range residuals for each two-day arc. These solutions were obtained without constraints, i.e., the a priori error covariance matrix was assumed to be infinite. The BIH five-day smoothed values with straight line segments constructed over each five-day interval are also shown in Figure 1 for reference. To assist in the interpretation of the laser results, a straight line ($a_o + a_1 t$, where a_o and a_1 are constants and t is time) was fit to the pole estimates using a weighted least-squares method. The weighting parameters were the variances associated with the estimates for x_p and y_p. These variances, which ranged from $(1 \text{ cm})^2$ to $(25 \text{ cm})^2$ for y_p, do not account for errors introduced into the estimates when modeling errors exist, but are necessary to represent uncertainties in the estimates due to differing geographical and temporal distribution of data. The variance weighted rms yielded 1.86 m for x_p and 1.68 m for y_p. When $\bar{C}_D$ and $\dot{\bar{C}}_D$ are estimated ($C_D = \bar{C}_D + \dot{\bar{C}}_D t$) the rms about the straight line fit was 1.86 m for x_p and 1.36 m for y_p. The straight line smoothing of the x_p estimates behave similarly to Figure 1; however, the fact that x_p is approximately perpendicular to the meridians of the laser stations suggests that the estimates will be sensitive to small changes. This is further reflected in the variances associated with the estimates which show that the x_p variance is usually greater than the y_p variance. Figure 2 shows the overall effect of drag by including the rms of an estimation in which drag is not modeled. Similarly, the drag exclusion has a significant effect on the x_p and y_p estimates. Further details on these results are given by Schutz, et al. (1978).

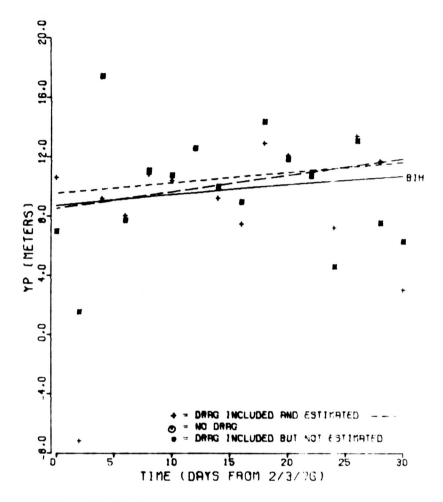

Figure 1. Estimates of y_p.

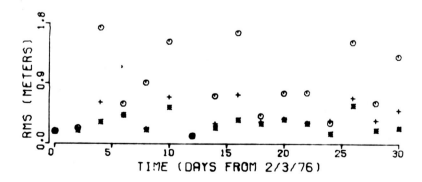

Figure 2. Rms of range residuals.

It is significant to note that results have been obtained using GEOS-3 data between 27 August 1975 and 10 September 1975 (Schutz, et al., 1978) which have also demonstrated the ability to estimate mean values of pole position which are in agreement with the BIH. For comparison, the results during this 1975 period produced a weighted rms of about 85 cm in the y component. The February results presented in this paper are especially significant since these data were not used in the GEM-10 geopotential and station coordinate solution, thus demonstrating an ability to predict the pole position which further reflects the overall model consistency.

There are two possible sources for the apparent systematic trends exhibited in Figure 1. Since the astronomical data provide no evidence for supporting the general behavior in this figure, most of the behavior must be attributed to unmodeled orbit perturbations. These perturbations exhibit frequencies similar to polar motion over the time period of their estimation, thus aliasing the x_p and y_p estimates. Further analysis was performed through use of the maximum entropy spectral analysis of Burg (1977). This analysis identified significant frequencies of about the same power at periods of 4.8^d, 5.5^d, 10.7^d, and 16^d. In addition, a 7^d period was identified with three times the power of the other frequencies. Since the resonance period of GEOS-3 due to fourteenth order terms in the geopotential is approximately 4.5^d, the 4.8^d period must be associated with "mismodeling" of these terms. Furthermore, the 10^d and 16^d periods can be correlated with ocean tide perturbations (Goad, 1977), leaving only the 5.5^d and 7^d periods of unknown origin.

Although a very significant earthquake occurred in Guatemala on 4 February 1976 at 9^h UT, there is no immediate correlation of the pole coordinates associated with the 760205 arc which significantly differ from the other values in February. Other earthquakes of similar magnitude occurred in this time period without comparable changes in pole position. Thus the observed behavior must be attributed to temporal and geographical data distribution.

CONCLUSIONS

These polar motion results have been obtained using regional tracking from three laser systems deployed off the U. S. East coast for GEOS-3 experiments. The results obtained have demonstrated the ability of the dynamical model and laser station coordinates to track the pole position with an uncertainty of less than two meters. This uncertainty indicates that additional model improvements are necessary for further analysis of GEOS-3 data. However, the results are significant in the implication that higher altitude satellites such as LAGEOS, which are substantially less dependent on model uncertainties from sources such as drag and geopotential, will provide polar motion estimates commensurate with the measurement precision.

ACKNOWLEDGEMENTS

This investigation was supported by the National Aeronautics and Space Administration through contract NAS 6-2454. The contributions of Mr. R. Eanes, T. Wilson, J. Lundberg, and C. Shum are gratefully acknowledged.

REFERENCES

Burg, J. P.: 1977, Paper presented at the 37th Ann. Int. Meeting Soc. of Explor. Geophys., Oklahoma City.
Dowd, D. L.: 1977, Institute for Advanced Study in Orbital Mechanics Tech. Rep. 77-4, The University of Texas at Austin.
Dunn, P. J., Smith, D. E.: 1977, J. Geophys. Res. 82, pp. 895-897.
Farrell, W. E.: 1972, Rev. of Geophysics and Space Physics 10, pp. 761-797.
Goad, C.: 1977, U. S. Department of Commerce, NOAA Technical Report NOS 71 NGS 6.
Lambeck, K.: 1971, Bulletin Geodesique, pp. 263-281.
Lerch, F., Klosko, S., Laubscher, R., and Wagner, C.: 1977, Goddard Space Flight Center Report X-921-77-246.
Schutz, B. E., Tapley, B. D., and Ries, J.: 1978, Institute for Advanced Study in Orbital Mechanics Tech. Rep. 78-4.
Smith, D. E., Kolenkiewicz, R., Dunn, P. J., Plotkin, H. H., Johnson, T. S.: 1972, Science 178, pp. 405-406.
Standish, E. M., Keesey, M., and Newhall, S.: 1976, Jet Propulsion Laboratory Tech. Rep. 32-1063.
Wilson, T.: 1978, Institute for Advanced Study in Orbital Mechanics Tech. Rep. 78-1.

DISCUSSION

H. G. Walter: Could you comment on your adjustment of the Love number which is significantly different from that of the Goddard Space Flight Center group?

B. E. Schutz: No Love number adjustment was performed. Instead, k_2, h_2, and l_2 were adopted from Farrell's Gutenberg-Bullen A Earth model. Furthermore, the results presented will not change at the decimeter level or less as the result of small perturbations in k_2, h_2, and l_2.

LOW COST LAGEOS RANGING SYSTEM

Peter L. Bender* and Eric C. Silverberg
Joint Institute for Laboratory Astrophysics
Boulder, Colorado 80309 USA
and
McDonald Observatory, University of Texas
Austin, Texas 78712 USA

ABSTRACT

This paper describes a Lageos laser ranging station design based on single photoelectron detection which could be used to monitor polar motion at all periods, and short-period UT1 variations with high accuracy and at moderately low cost. No comparison has been made with the costs of long baseline radio interferometry (LBI) stations, since the establishment and operation of basic international networks using both approaches seems scientifically well justified. Both LBI and lunar laser ranging results will determine intermediate and long-period UT1 variations and nutation. The LAGEOS laser ranging station design suggested here is similar to the design used in the high-mobility LAGEOS ranging station now being constructed by the University of Texas. It has been shown that 30 cm diameter transmit/receive optics with 10 arc second pointing accuracy and 50 milliwatt average laser power output are usually sufficient to give 70 signal counts in a three minute averaging time, even at 20^0 elevation angle. Lasers having the desired power at 530 nm wavelength with 10 pps repetition rate and 3 ns pulse length are now reported to be commercially available at moderate cost. The resulting range accuracy for normal points corresponding to the three minute averaging time is three cm, including allowances for absolute calibration of the system, atmospheric refraction correction uncertainties, and Earth tide uncertainties. Much of the pulse-timing electronics needed is commercially available, and many time and latitude stations already have good epoch determination systems. The ten arc second pointing requirement for a 30 cm diameter beam seems considerably easier to meet than the optical requirements for classical observing techniques. A simple beam director with flat mirrors appears quite feasible, with inexpensive encoders on each axis for minicomputer control if desired. A number of special features of the University of Texas high-mobility station would not be needed for a fixed station.

*Staff Member, Quantum Physics Div., National Bureau of Standards.

DISCUSSION

Ya. S. Yatskiv: Why have you changed your plans for the construction of a mobile lunar ranging station?

E. C. Silverberg: The increased possibility of aseismic slip in non-earthquake zones favors the measurement of many locations on each continental plate. After much soul searching, it was decided to defer the development of the transportable lunar station in favor of a more mobile system optimized for Lageos. The transportable lunar station will be slowly completed over the next few years, to act ultimately as a stand-alone multipurpose replacement for the McDonald Observatory 2.7m facility.

PART VI : LUNAR LASER RANGING

ON THE EFFECTIVE USE OF LUNAR RANGING FOR THE DETERMINATION
OF THE EARTH'S ORIENTATION

Eric C. Silverberg
Department of Astronomy
University of Texas
Austin, Texas

ABSTRACT

Lunar ranging data have been routinely available since September of 1970, but many problems of a varying nature have delayed the establishment of a world-wide lunar ranging network. As a result, we must re-examine the role which this program can play in the determination of the Earth's rotation and polar motion. Although there are many technical difficulties now inhibiting the widespread use of this technique there seems little doubt but that we can overcome these problems and achieve routine, accurate orientation determinations. The more difficult questions concern how an Earth rotation campaign should now be configured to use the equipment and resources in the best way.

Despite considerable progress by other techniques, the failure to develop a lunar capability for Earth orientation determinations would result in a serious loss of information. Lunar monitoring of long-term effects in the Earth's rotation rate and the relationship of the lunar orbital parameters to a stellar reference frame are two tasks for which there is little redundancy. However, the most cost-effective usage of station resources may not require daily measures, but only periodic, accurate snapshots of the Earth's rotational position relative to the Moon. If some satellite laser ranging facilities could treat the Moon as an object of opportunity, but were able to elevate the program to priority status when global conditions were favorable, the incremental cost of gathering the required lunar data might be drastically reduced. Even though these cost savings could not be achieved without daily communication between cooperating stations, such a detailed interaction is not unreasonable to consider.

1. INTRODUCTION

At this writing, no less than nine laser ranging stations in five different countries have announced returns from retroreflectors deposited on the lunar surface by the U.S. and Soviet space programs (Bender, et al., 1973). The typical lunar laser ranging system fires a short pulse of visible light containing approximately 10^{18} photons. If the beam is highly collimated, such that it only illuminates a few square kilometers of the lunar surface, the many-cornered retroreflectors will return sufficient energy to cause a detectable signal at an Earth-based receiver. The basic data recorded by the lunar ranging system consists of the epoch at which the laser shot was fired, the round-trip travel time to the lunar surface, the environmental parameters, and the system calibration corrections.

What is, in principle, a very simple process and merely an extension of artificial satellite ranging technology has proven sufficiently difficult that only McDonald Observatory, of the original nine stations, has maintained consistent data over the last eight years. Many of the problems which have prevented regular observations on a more widespread basis are related to financial constraints and the normal delays to be expected in activating any new technology. Another set of problems, however, is more serious and relates to physical constraints or inherent weaknesses which will remain indefinitely. The purpose of this document is to review our history with the lunar ranging technique, explore some of the reasons which have limited the acquisition of the basic data, and to propose a set of goals based on this nine-year experience which addresses the role which this type of data will play in conjunction with other available techniques.

2. THE LIMITATIONS

Lunar ranging, by its very nature, works close to the threshold of detection for current laser radar technology. If, for instance, the Moon were 30 percent further away, no lunar ranging system currently in existence would be likely to attain routine operations. This peculiar circumstance can be attributed to the state of development in the laser itself. Key parameters which govern the design of the laser radar system are laser pulse width, average power, and the laser divergence. Early lasers required large aperture instruments, such as the 2.7-meter McDonald reflector, due to their large intrinsic divergence. Great strides in the improvement of this area have made practical the use of smaller telescopes, a fortunate occurrence since few of these larger instruments would be available for such an experiment. Shorter pulse width coupled with higher efficiency optics and better detectors have been able to continually improve accuracy despite the lack of collecting power in the smaller apertures. However, there have been no average power gains commensurate with the improvement in the other parameters. The 1.5 watt capability of the McDonald laser is still well within an order of magnitude of the state of the art nine years later. Thus, unless there is a major breakthrough in the availability of high average powered lasers, the experiment will forever require highly collimated output beams and be at the mercy of atmospheric turbulence as well as clouds. Moreover, the high beam collimation required for the lunar work dictates a demanding pointing technology in comparison with other techniques (Silverberg, 1974).

Any operation which operates close to the threshold of detection presents a practical problem for the development of new facilities. The accompanying financial threshold is a very real problem, in that stepwise turn-on is not possible like in many other experiments. A considerable capital commitment is necessary before demonstrable results can be obtained. The same is true in a technical sense. A lunar ranging system is a relatively complex assembly of components, any one of which can prevent the acquisition of data -- not only prevent the acquisition of merely good data, but prevent the acquisition of any data. Once quasi-regular data is acquired from a lunar ranging station, it should be relatively easy to iterate the system to a routine astrometric operation. Unfortunately, that first step is more difficult than we hoped.

3. THE ADVANTAGES

There are, of course, also advantages to lunar ranging which have led many investigators to put a great deal of effort into developing ranging facilities throughout the world. The first and foremost reason for pursuing lunar ranging, as opposed to some other method, is the unique access to the Moon's dynamics. The twofold improvement in monitoring the Moon's motion allows the study of general relativistic terms, the various harmonics of the Moon's rotation, the mass of the Earth/Moon system, and many other aspects which are not studied elsewhere (Williams, 1977). In the sense that the Moon's orbit is very important to Earth rotation, these studies alone are of interest to a symposium of this nature. However, the main reason for discussing this technique in the context of Earth rotation is the sensitivity of any lunar laser station to the time of lunar meridian passage. Each lunar ranging system can, given a successful set of three observations in a single day. determine the time of meridian passage and the geocentric axial distance to approximately the same accuracy as the measurements. A large number of determinations of UT0 have been possible for the McDonald station over the last several years with a formal accuracy of about 0.7 milliseconds, although differences between various analyses indicate that the actual error must be somewhat greater (Stolz, et al., 1976; King, et al., 1978; Harris and Williams, 1977; and Shelus, et al., 1977). Several stations can cooperate to use either the axial distance determinations to orient the Earth's pole, or can directly infer the orientation geometrically once the lunar declination is well known. This monitoring of geocentric station position relative to an inertial reference frame provided by the solar system is a unique contribution in this area, important to tracking the long-term variations in Earth dynamics. Moreover, the analysis of lunar data for rotation variations appears to be much simpler than demonstrated for other techniques, so simple in fact that the on-site ranging crew has a real sense of the status of UT0. Thus, while lunar ranging appears to have the most difficult data gathering problem, this drawback is offset by the simplicity and economics of the data analysis. The fundamental reference provided

by the lunar orbit and the simplicity of the data analysis led to
strong hopes that an operational system for Earth rotation would be
highly competitive with other techniques, even for short-period studies.

4. CURRENT STATUS

The McDonald Observatory station has continued to operate regularly
since September of 1970. The normal operating schedule, weather permitting, is three trials per day, approximately 300 days per year. The
precision of the measurements is normally about 10 cm, however 5-cm
ranges are possible under excellent conditions. Although the development of the station was recently slowed somewhat by financial constraints, the installation has now returned to its regular operating
schedule. McDonald has collected over 2300 range measurements to four
of the lunar corner reflectors during the last seven years. To this
author's knowledge, only the 2.6-meter installation at the Crimean Astrophysical Observatory has also been in regular operation during this
period, conducting a limited set of observations which averages a few
days per lunation.

Two other lunar laser ranging stations have all of the equipment presently in place and have made some preliminary observations of range.
Unfortunately, neither has reached operational status at this writing.
The Maui station was an ambitious attempt by the U.S. lunar ranging
team to achieve a second generation system which would have not only
2-cm accuracy and 3 times the McDonald signal level, but have the ability to perform blind tracking during day and night-time conditions.
Unlike most other experience, the major problem at the Maui station
is not one of signal strength, which has been excellent when obtained,
but the high complexity of the system which requires extreme reliability in all components and a high degree of crew training.

The Australian station near Canberra is designed around a 1.5-meter
telescope formerly used by the AFCRL organization near Tucson, Arizona. This station has been conducting preliminary ranging operations
for approximately a year. While there are some difficulties with reliability of components, particularly the laser, the major problem for
this station has been the poor signal-to-noise ratio caused by the
wide laser pulse and the necessity for large spatial filters.

These four installations can be joined by as many as three others in
the near future. A second generation CERGA installation at Grasse is
now nearing completion with a 1.5-meter telescope. Modifications are
underway to upgrade the artificial satellite station at Wettzell, and
it is known that the Tokyo Astronomical Observatory intends to begin a
program of lunar range observations.

If any substantial subset of the possible lunar stations can begin coordinated operations, it seems likely that high quality Earth rotation
and orientation data will be attained, as envisioned by the COSPAR

sponsored EROLD campaign. Modelling by Stolz and Larden (1977) indicate that a regular lunar ranging capability can effectively track the expected variations in pole position with normal weather constraints. Data handling and analysis methods for an Earth rotation service have been demonstrated in coordiantion with the BIH. There is no doubt that an ambitious program of state-of-the-art Earth rotation measurements is possible if the current station problems can be overcome. The question which should be asked, however, is whether the ambitiousness of these goals is, in fact, the reason why progress has been slow. This question is especially relevant when it is possible to foresee a much less ambitious observation program which still contributes substantially to the scientific needs in these areas and provides the necessary information from which further plans could be drawn.

5. COMMENTS ON THE GOALS

The difficulty of designing and operating a lunar laser ranging station is highly related to the percentage of the lunation which the station intends to cover. The necessity for full month operation requires not only a variety in guiding techniques, but the ability to operate day and night during substandard seeing and marginal contrast conditions. Table 1 attempts to characterize the varying degrees of difficulty in relation to the operating time in each lunation, related to items such as the initial complexity of the equipment, the technical operating expertise, and the cost of the operation. With an operating time of five or six days per lunation, the equipment requirements mimic a slightly modified artificial satellite station. On the other hand, twenty-eight day coverage of the orbit requires a level of expertise which has still not, to this investigator's knowledge, been demonstrated. The expense in operating the station must also rise with the ambitiousness of the program, with the first break in cost coming approximately at 12 days per lunation when one crew can no longer be expected to perform all of the functions associated with operating and maintaining such a facility. It is evident that all aspects are highly correlated with the percentage of operating time which is envisioned for each lunation. Good system design will recognize this fact by coordinating the initial investment in equipment with the long-term operating plans for the station. It is also important, however, that the worldwide ranging community recognize this fact with goals which are commensurate with the status of the facilities as they now exist.

Table 2 lists potential goals for the lunar laser ranging technique and the observational requirements, both in order of increasing capability of the stations to cover the entire lunation. Goals which are unique to lunar laser ranging have been indicated with an asterisk. Most require relatively small numbers of observations from a limited number of lunar laser ranging stations. Very important worthwhile projects can be accomplished if a few laser ranging stations have occasional lunar capability. On the other hand, the most ambitious goal for lunar laser ranging, that is, the development of an independent capa-

Table 1. Characteristics related to the percentage coverage of the lunation.

Days Per Lunation	Initial Investment	Operating Technology	Operating Costs Devoted Lunar	Operating Costs Combined with Satellite
28	impractical	---	---	---
25	very high	very high		
20			2 crews	moderate
15	moderately high	routine	1.5 crews	
10			1 crew	low
5	moderate	low		
0				

bility for daily Earth orientation determinations (a need not demonstrated), can also be made by at least two other methods. While lunar ranging can be an effective and unique contributor in the area of Earth rotation and polar motion, it does not appear that the pursuit of daily independent determinations of the latter are justified at this time.

6. RECOMMENDATIONS

In the area of Earth rotation, the most cost-effective goals are the maintenance of the secular variations. Accurate snapshots of the Earth's orientation can be economically achieved by a few laser stations conducting quasi-simultaneous observations for a few days per month. With daily polar tracking by other techniques, any single station's observations of the Moon during the month will also track the Earth's short period rotation variations. Thus, occasional lunar tracking by several stations and frequent lunar tracking by one or two stations will deliver the greatest percentage of the scientific gains, lacking only a daily independent Earth orientation solution. All of the unique lunar scientific goals would be attainable with such data. These suggestions are attainable, cost-effective programs which do not require full lunation capability by a large number of lunar tracking facilities.

Without the motivation associated with the daily independent service for the determination of Earth orientation, it is likely that some of the support for lunar ranging that might otherwise be available will be lost. There is a definite risk that so much support could be lost that even the unique goals available to the data type would be jeopardized. Hopefully, this problem can be successfully addressed by realizing that limited observational goals for a lunar ranging station make it more compatible with other systems. Lunar ranging systems which are designed for a relatively small portion of the lunation need not grossly differ from a satellite ranging configuration. Moreover, the observations can be intermixed with little additional personnel cost

Table 2. Observing requirements.

GOAL	REQUIRED CHARACTERISTIC FREQUENCY OF OBSERVATIONS	REQUIRES MULTIPLE RETROREFLECTORS	REQUIRES MULTIPLE STATIONS WITH EARTH ROTATION/POLE SERVICE	REQUIRES MULTIPLE STATIONS WITHOUT EARTH ROTATION/POLE SERVICE
* Lunar Orbital Studies: gravitation, tidal, fundamental const., relativity	weekly	some	no	probably
* Lunar Libration Studies: selenodesy, c/mr^2, free librations, Q	weekly	definitely	no	probably
Establish Geodetic Benchmarks: continental drift, fault monitoring	monthly	no	yes	yes
Monitor Secular Variations In Earth Orientation: rotation vs. solar system*, pole wander	monthly	no	yes	yes
Provide "Daily" Earth Rotation Service	daily	no	no	yes
Provide "Daily" Pole Motion Service	daily	no	yes	yes

* Unique to Lunar Ranging Techniques

at the same site whether or not the actual hardware is shared between the two systems. It can only be advantageous to have multiple data from the same location when pursuing difficult comparisons of geodynamic solutions at the centimeter level. It seems evident that a very cost-effective package could be addressed to funding agencies in terms of multiple purpose laser facilities, capable of ranging to the Moon as well as a wide variety of lower objects. These stations could track artificial satellites for much of the month, but would be available periodically to make high accuracy snapshots of the Earth's orientation relative to the Moon, hence the solar system, at favorable times. The observations by the multiple stations could easily be coordinated such that they occurred on or near the same day, once the capabilities of all the stations were determined.

7. SUMMARY

Surveying the situation for lunar ranging in the context of this symposium, we have, on one hand, the observational difficulties associated with obtaining data, the gap at new Moon, and the lack of many observatories. On the other side are the unique contributions in the field of lunar science, the geocentric measurements in an inertial frame and the very simple and economical data handling and analysis. The establishment of an effective daily independent determination of Earth orientation does not seem cost-effective at this time, due to the requirements for coverage of the entire lunar cycle implied by this activity. On the other hand, the foreseeable laser network can expect to track longer term (monthly) variations in Earth rotation and polar motion in a cost-effective manner, particularly if further consolidation with satellite tracking problems is achieved. Lastly, the availability of polar motion data from another technique would permit the very economical, real time determination of Earth rotation from any individual lunar station which happened to operate over a high percentage of the lunation.

These discussions have been limited to questions relating to the basic data gathering activities of lunar ranging. It must be emphasized that this alone is not the total picture. The final measure of the success of any technique must be based on many factors including proven accuracy and relative costs. It has been proposed that, in view of the difficulties which have occurred in the development of lunar laser ranging stations, less ambitious immediate goals should be considered for lunar laser ranging than those envisioned by the international EROLD campaign -- lesser goals which can be chosen without drastically altering our access to unique scientific data. The final decisions between what should or should not be considered long-term goals for this technique are still open questions and will be for some time. However, it seems likely that by combining the lunar laser ranging equipment, to as large an extent as possible, with the need for tracking artificial satellites, an extremely cost-effective package can be developed. The use of further coordination with the pole solutions from other areas such as VLBI can further increase availability of scientifically meaningful data from individual lunar stations. It is evident that while lunar ranging has much to offer, it's most cost-effective scenario includes detailed coordination with complementary methods. It is of interest to consider whether the same is true of the other proposed techniques.

This work is supported by NASA Grant NGR 44-012-165.

REFERENCES

Bender, P. L., Currie, D. G., Dicke, R. H., Eckhardt, D. H., Faller, J. E., Kaula, W. M., Mulholland, J. D., Plotkin, H. H., Poultney, S. K., Silverberg, E. C., Wilkinson, D. T., Williams, J. G., Alley, C. O.: 1973, Science 182, p. 228.
Harris, A. W. and Williams, J. G.: 1977, in J. D. Mulholland (ed.), "Scientific Applications of Lunar Laser Ranging", D. Reidel, Dordrecht, p. 179.
King, R. W., et al.: 1978, J. Geophys. Res., (in press).
Shelus, P. J., Evans, S. W., and Mulholland, J. D.: 1977, in J. D. Mulholland (ed.), "Scientific Applications of Lunar Laser Ranging", D. Reidel, Dordrecht, p. 191.
Silverberg, E. C.: 1974, Applied Optics 13, p. 565.
Stolz, A. and Larden, D.: 1977, in J. D. Mulholland (ed.), "Scientific Applications of Lunar Laser Ranging", D. Reidel, Dordrecht, p. 201.
Stolz, A., Bender, P. L., Fuller, J. E., Silverberg, E. C., Mulholland, J. D., Shelus, P. J., Williams, J. G., Carter, W. E., Currie, D. G., and Kaula, W. M.: 1976, Science 193, p. 997.
Williams, J. G.: 1977, in J. D. Mulholland (ed.), "Scientific Applications of Lunar Laser Ranging", D. Reidel, Dordrecht, p. 37.

DISCUSSION

W. E. Carter: Lunar ranging near new Moon is difficult, and measurements of Earth rotation could only be obtained on about 20 days per month unless sophisticated systems were used. This means that lunar laser ranging is not, by itself, a suitable technique for an Earth rotation monitoring system.
E. C. Silverberg: That is true.

IS LUNAR RANGING A VIABLE COMPONENT IN A NEXT-GENERATION
EARTH ROTATION SERVICE?

J. Derral Mulholland
McDonald Observatory & Department of Astronomy
University of Texas at Austin, USA

ABSTRACT

Several new "space" techniques have been used for episodic determination of Earth rotation parameters, usually the variation in apparent longitude (UT0) and apparent latitude of an observing station. Earth rotation services require more than episodic determinations; they need near-daily determinations. Since 1975, planning has been underway for a demonstration of the viability of lunar laser ranging for such a usage. The observing campaign named Earth Rotation from Lunar Distances (EROLD) was organized with the proposed activity to cover the years 1977-78. Progress has not been so rapid as hoped, but it remains true that lunar ranging has produced more Earth rotation information than other new techniques.

Silverberg (1979) has dealt with the generalities of the application of lunar laser ranging (LLR) to the problem of the experimental determination of Earth rotation. The theoretical bases have been so thoroughly presented in previous symposia and Commission meetings (e.g. Mulholland, 1977) that it seems unnecessary to repeat them here. It is rather my purpose to discuss the present actuality. In doing so, I will begin by stating my prejudice: Earth rotation is a phenomenon that requires continual monitoring, and any system that proposes determination of Earth rotation must address itself to the hard questions of service bureau operations before it can be taken seriously.

It was in this spirit that, in 1974, a proposal was presented to COSPAR for an observing campaign to be called "Earth Rotation from Lunar Distances" (EROLD). The goal was not just to make episodic determinations of the Earth rotation parameters, but to try to meet a regular and frequent schedule, such as those now followed by the BIH and the IPMS networks. The idea met with approval, and the organization moved forward, with the goal of a coordinated observing campaign to begin on 1 January 1977. The Bureau International de l'Heure accepted the responsibility for the combined data reduction, in collaboration with the Centre d'Etudes et de Recherches Géodynamiques et Astronomiques; the EROLD Steering Committee was composed with representatives of the two international services and each of the national observing groups. The date for beginning the campaign was chosen because it seemed likely that at that time there would be at least three stations that would be fully operational at that time; this is a practical (though not a theoretical) minimum for such an operation. As noted by Silverberg, however, this is a technique in which the relative ease of data reduction and analysis is heavily paid for in the difficulty of the operational aspect. The result is that, in May 1978, there is still only one station producing regular data of adequately high quality. The BIH/CERGA data processing system has been ready for experimental use since the beginning of 1977; in the absence of multiple stations, it has been used only to determine UT0 for the McDonald Observatory (Calame, 1979). Perhaps it is worthwhile here to give a brief summary of the observing situation.

AUSTRALIA -- The station at Orroral recorded its first statistically-significant event in July 1977. Since then, several system problems have been discovered and solved. It is believed that high-quality observations could begin any day.

FRANCE -- The new 1.5 m telescope was installed at Calern in June 1977. After the replacement of faulty parts, tracking tests were resumed in this past month (April 1978). The refurbished laser, which had been operated successfully in the Pic-du-Midi station, is installed, as is most of the other equipment. Operations could begin later this year.

GERMANY (Federal Republic) -- The artificial satellite laser station at Wettzell is being upgraded to lunar capability, but the first attempts at the Moon cannot be anticipated before Spring 1979.

JAPAN -- The Dodaira station has experienced much hardware difficulty. Recent modifications permit one to hope for rapid progress.

USA -- The McDonald station, in continuous operation since July 1969, continues to show that the technique is feasible. Although operations were reduced to 18 days per lunation for several months due to budget restrictions, they are back to 21-day operation now. For several years, McDonald has averaged about 25-30 observation normal points per lunation. Accuracies of 5-cm are now common, although not yet the rule. The Haleakala station continues to have a detector problem that prevents their lunar echos from being timed with an accuracy better than about 5 ns (75 cm). Regular high-quality data should be acquired regularly as soon as this problem is identified and fixed.

USSR -- Ranging operations at the Crimean station are still conducted only 20-25 days per year, which is inadequate for Earth rotation use. A dedicated transportable station is expected to be ready in 1980.

Thus, a summary of the present situation of EROLD is:
-- The data analysis and distribution is ready;

-- One station demonstrates that near-daily operations can be maintained;
-- Three-station operation may become a reality at any moment;
-- Full network (5-6 station) operations cannot be expected before late 1978 or early 1979.

Unfortunately, excepting dates, this summary has not changed in more than a year. With only one station, it has not been possible to measure UT1 or the coordinates of the pole. On the other hand, it is not true that LLR has produced negligible Earth rotation information. In addition to Calame's work already cited, the following UT0 determinations have been made:

-- Stolz et al (1976) presented values obtained on 194 separate days over the interval 1970-75, applying a very strict hour-angle criterion to select the days to be studied;
-- Shelus et al (1977) selected a specific lunation, using all of the data to obtain both daily and 2-day means nearly continuously over a 21-day time interval;
-- King et al (1978) have analyzed essentially all of the data from 1969-76, but with averaging times of up to ten days, giving on the order of 150-200 determined values;
-- Williams (1978) reports having obtained UT0 values for about 500 individual days over the interval 1970-77.

The density is not high enough, but it is much better than is generally realized. In fact, there have been more determinations of UT0 by lunar ranging than by all of the other "space" techniques combined, and it is worse than an error to overlook this fact. The method has the unique advantage of providing a direct tie to the celestial coordinate system with only one station per observation; there are other advantages, and even disadvantages, but these have been detailed elsewhere (e.g. Mulholland and Calame, 1977).

I do not wish to leave you with the feeling that I know or believe that LLR is the solution for an Earth rotation system of the future. In fact, I believe that the future system (like the present one) will combine several techniques, in an attempt to capitalize on the strong points of each. The purpose of EROLD is to test a specific technique in a real-world mode, so that the strong and weak points might be evaluated realistically. It is a procedure that I recommend for each of the new techniques. When that has been realized, then rational choices can and (I hope) will be made.

ACKNOWLEDGEMENTS

I am grateful to E. C. Silverberg and O. Calame for helpful discussions and criticism of the manuscript. Preparation of this paper was supported by the U. S. National Aeronautics and Space Administration, under grant NGR 44-012-165.

REFERENCES

Calame, O.: 1979, this volume.
King, R. W. et al: 1978, J. Geophys. Res. (in press); abstract 1977, in J. D. Mulholland (ed.), "Scientific Applications of Lunar Laser Ranging", D. Reidel, Dordrecht, p. 219.
Mulholland, J. D.: 1977, in J. D. Mulholland (ed.), "Scientific Applications of Lunar Laser Ranging", D. Reidel, Dordrecht, p. 9.
Mulholland, J. D. and Calame, O.: 1977, in L. J. Rueger (ed.), "Proceedings 9th Precise Time and Time Interval Planning Meeting", NASA Goddard Space Flight Center, Greenbelt.
Shelus, P. J., Evans, S. W. and Mulholland, J. D.: 1977, in J. D. Mulholland (ed.), "Scientific Applications of Lunar Laser Ranging", D. Reidel, Dordrecht, p. 191.
Silverberg, E. C.: 1979, this volume.
Stolz, A. et al: 1976, Science 193, p. 997.
Williams, J. G.: 1978, private communication.

DISCUSSION

P. Paquet: In suggesting that lunar ranging should be a component of a future Earth rotation monitoring service, do you feel that it should contribute on a daily basis or as a system to recalibrate the reference system used to define the orbit of Lageos?

J. D. Mulholland: Whether lunar observations are required on 5, 10 or 20 days per month is, I think, still to be determined from experience. In any case, LLR will be required for Lageos calibration, because of the problem of rectification already mentioned by Melbourne. An orbit rectification is an artificial discontinuity in the mathematical description of the orbit, and it is needed because the models of the acceleration are inadequate. The Moon is free from this problem; when we speak of data analysis over an interval of eight years we mean a continuous, dynamically consistent orbit over the entire interval. This is important for geophysical reasons. For example, if the Chandler motion is driven by seismic activity then large seismic events will introduce real discontinuities into the rate of polar motion and rotation. It is important that such phenomena be discussed with techniques from from artificial discontinuities that could mask or mimic the real effects. Thus, it seems to me that significant LLR activity will be needed, but that the density of observations required is not yet known. This should be a primary goal of experimental observing campaigns such as EROLD.

PRELIMINARY UT0 RESULTS FROM EROLD DATA

Odile Calame
BIH/CERGA
Ave. Copernic
06130 Grasse, France

INTRODUCTION

The status of the EROLD campaign has already been described in another paper (Mulholland, 1979). Since the beginning of 1977 all of the computer software has been ready to work. With the expectation that at least one other station would be able to operate before the end of 1977, it appeared not very interesting to use the single-station data for the study of UT0 during the last year since the other parameters of the Earth's rotation can be determined only with multiple stations. However, as this situation of a single operating station continues, it seemed to be preferable to attempt some determinations of UT0, even if that represents only a part of the goals envisioned from this campaign. We give here a brief resume of our preliminary results.

DETERMINATION OF UT0

The determinations of UT0 have been made for the available data from the McDonald station, using four lunar reflectors. Except for some improvements, the mathematical model used to compute the residuals and partial derivatives has already been described elsewhere (Calame, 1976); recently the effects of tides for an elastic Earth model have been introduced.

Several determinations have been made for UT0 differing by the selection of the data and the time intervals for which each value of UT0 was determined. For example, averaging has been performed over 5 days, 2 days, and 1 day with the constraints that the extreme hour angles in each interval differ by at least 3 hours and that the number of data be greater than 5 to 10. A first fitting pass consisted of an adjustment for 46 parameters describing the motions of the reflectors with respect to the center of mass of the Earth. In addition, some unknowns for the geocentric position of the station were included: coordinates, drift in longitude, annual term in longitude, and a long-period term (18.6 years). In a second step, new post-fit residuals were computed and another adjustment was made to take into consideration the geocentric unknowns for the station (mentioned above) and corrections to UT0 taken from the 5-day smoothed values published by the BIH and used in the residual computations.

Figure 1 shows a sample of the results obtained for UT0 from 2-day determinations for the year 1977. For the moment, it appears to be difficult to detect any short-period effects with confidence. The rms residual in the time interval 1970 - 1978 is about 36 centimeters.

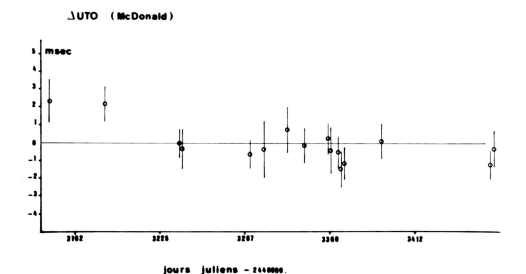

Figure 1. Corrections to BIH UT0 determined from EROLD data.

CONCLUSION

It is necessary to emphasize that these results concern only UT0 determinations. It would be unrealistic to compare them with UT1 determinations obtained in other ways. Indeed, some local effects may perturb the results for a single station so that the global Earth rotation may differ from them. Only when other stations become operational will it be possible to study this question along with polar wander. As soon as data from other stations will be available we will be able to get new results very quickly, for example, in one or two weeks.

REFERENCES

Calame, O.: 1976, Manuscripta Geodaetica, vol. 1, no. 3.
Mulholland, J. D.: 1979, this volume.

PART VII : DOPPLER SATELLITE METHODS

POLAR MOTION THROUGH 1977 FROM DOPPLER SATELLITE OBSERVATIONS

Claus Oesterwinter
Naval Surface Weapons Center
Dahlgren, Virginia

ABSTRACT

Doppler observations of Navy Navigation Satellites have been used to compute pole positions on a daily basis since 1969. Limited results exist for the period 1964 to 1969. Based on Doppler observations from four or five satellites, the standard error for a five-day mean pole position is less than 20 cm. Comparisons are made between BIH, IPMS and ILS results and those obtained from Doppler. It is shown that the six years of reliable Doppler data since 1972 contribute little in finding the Chandler period. Using observations from the three astronomical sources over 12 years yields a Chandler period of 432.0 ± 0.2 days.

PREVIOUS WORK

The determination of the coordinates of the Earth's spin axis from Doppler observation has been described by Anderle and his colleagues in a series of publications and reports. The method of computation was briefly explained by Anderle and Beuglass (1970). A more detailed description of the observational procedures, the data reduction techniques, and error sources was given by Anderle (1973a).

Results of Doppler data analysis and comparison with other determinations are discussed in numerous places. All Doppler results are based on five-day mean values of x and y. They are tabulated and discussed for 1969 by Anderle and Beuglass (1970), for 1967 to 1970 by Anderle (1970), for 1969 to 1971 by Anderle (1972), for 1972 by Anderle (1973b), and for 1973 by Beuglass (1974). The five-day means for the years 1974 to 1977 are given in this paper. A few two-day solutions for 1964 to 1969 may be found in Anderle (1973b), Appendix F.

The above data are normally shown as plots of x vs time, y vs time, and x vs y. They are given by Anderle and Beuglass (1970) for 1969, by Anderle (1970) for 1967, 1968, 1969 and 1970, by Beuglass and Anderle (1972) for 1970, by Anderle (1972) for 1969, 1970 and 1971, by Anderle

(1973b) for the period 1964 through 1967 and for 1972, by Anderle (1973a) for mid-1971 to mid-1972, by Beuglass (1974) for 1973, and by Anderle (1976b) for 1975. Plots for the years 1974 to 1977 are shown in this report.

Anderle (1976a) has also compared Doppler derived pole coordinates with classical optical solutions. He plots the differences BIH-ILS, DMA (Doppler) - BIH, and DMA-ILS for the span 1964 to 1975. He also tabulates yearly mean values for above differences as well as associated statistics. Anderle (1976b) adds the comparison DMA-IPMS and shows more detail by breaking the plots into two spans, 1964 through 1969 and 1970 to 1975.

OBSERVATIONS AND DATA REDUCTIONS

The following is a very brief description of observational data and their analysis. Details may be found in the references listed at the end of this report, especially in Anderle (1973a) and Anderle (1976b).

Observations are the Doppler shifts in the continuous radio frequencies at 150 and 400 MHz transmitted by the U. S. Navigation System satellites (Kershner, 1967). Analog combination of these two frequencies permits elimination of one large error source, namely the first order ionospheric refraction effect.

The number of satellites being observed varies between two and five depending on Navy requirements. Table 1 shows which satellites were observed, and when, for the years 1974 to 1977.

Table 1. Available Doppler satellite data (Day Numbers).

	1967-34A	1967-48A	1967-92A	1970-67A	1973-81A
1974		166-280	1-87	89-363	
1975				2-362	13-363
1976	155-365			6-364	1-157
1977	7-167	21-167	21-167	6-364	21-365

Observations are taken by as many ground stations as are operational. They increased in number from about 13 in 1969 (Anderle and Beuglass, 1970) to about 20 in recent years (Anderle, 1976b).

All observations taken during a 48-hour time span are used in a least squares solution to improve, primarily, the orbital parameters. During this process, the satellite orbits are numerically integrated by Cowell's method, that is, the Gauss-Jackson algorithm spplied to the differential equations in the rectangular accelerations. The program is normally run with a 60 seconds integration step size and order 12. The reference frame is the mean equator and equinox at the beginning of the observation span. The mathematical model contains about 480 gravity terms, atmospheric drag, radiation pressure, luni-solar solid Earth tides, with the Love number presently set at 0.26. The force field is complete enough to permit determination of the satellite's position good to about one meter.

The solution also contains, among other parameters, the coordinates x and y of the spin axis, referred to the CIO. Such two-day solutions are obtained separately for each satellite. Subsequently all two-day solutions from up to five different satellites are combined into five-day means. The latter are published by the U. S. Naval Observatory in "Preliminary Times and Coordinates of the Pole, Series 7."

The computation of pole positions based on Doppler observations originated at the Naval Weapons Laboratory, now the Naval Surface Weapons Center (NSWC). In April 1975 the responsibility of computing NAVSAT orbits, and, hence, the derivation of pole positions, was transferred to the Topographic Center of the Defense Mapping Agency (DMA). Since DMA employs the same computer programs, the transfer did not affect position results.

Over the years, there have been a number of changes in the observation station network and observation techniques (Anderle, 1973a) as well as improvements in the data reduction methods (Anderle, 1972). However, the procedures have been essentially the same since August 1971, so that Doppler results after this date are believed to be homogeneous.

DOPPLER POLE POSITIONING ACCURACY

The formal standard deviation for the polar coordinates from a two-day solution is about 5 cm during the second half of 1977. But it must be remembered that such solutions are made for each satellite separately. All two-day solutions are then combined into five-day means. Subsequently, one can compute the more realistic standard deviation of a two-day coordinate with respect to the five-day mean. That number is presently a bit less than 40 cm. The standard deviation of the five-day mean itself (standard error) has been just under 20 cm for the last two years.

The increase in accuracy from 1967 to 1977 is shown in Table 2. However, the data before and after 1972 are not immediately comparable. Polar coordinates until August 1971 were extracted indirectly from called-for corrections to station coordinates, and they were one-day solutions. Moreover, they were computed after orbit improvement, not in a simultaneous least-squares solution.

Table 2. Preliminary yearly rms of Polar Coordinates.

	Standard Deviations (two-day solutions *)			Standard Errors (five-day means)		
	x [m]	y [m]	Av. [m]	x [m]	y [m]	Av. [m]
1967	1.65	1.78	1.72	.89	.74	.82
1968	1.48	1.60	1.54	.86	.93	.90
1969	1.51	1.27	1.40	.69	.60	.65
1970	1.25	1.15	1.20	.57	.53	.55
1971	1.16	1.39	1.28	.52	.62	.57
1972	.75	.69	.72	.37	.32	.35
1973	.38	.44	.41	.22	.28	.25
1974	.48	.50	.49	.30	.32	.31
1975	.47	.36	.42	.22	.18	.20
1976	.40	.30	.35	.20	.15	.18
1977	.41	.34	.38	.18	.15	.17

*One-day solutions before 1972

Anderle (1973a) pointed out that the principal error source in Doppler polar coordinates is due to inadequate knowledge of the Earth's gravity field. Despite recent advances, this remains true today.

RESULTS 1974-1977

Anderle and his colleagues have already published diagrams and tables summarizing polar motion during 1974 and 1975. Since some of their results were based on preliminary data, they are repeated here using final data. Final values were also available for 1976 while some 1977 results are still preliminary. They will be identified as such below.

Tabulation of Doppler Results

Table 3 is a sample containing the two-day and five-day Doppler solutions for polar coordinates. The complete tables for the years 1974 to 1977 will be published in a forthcoming NSWC report. The first two columns show the day numbers for each two-day solution. They are followed by x and y and their formal standard deviations (labelled "Standard Error") as obtained from the covariance matrix of the least-squares solution. The last two columns are the satellite designation and the nominal value for UTC-UT1. The latter information is not used in our pole position calculations.

Table 3. Sample Doppler solutions for polar coordinates.

DAHLGREN POLAR MONITORING SERVICE
NWL 9 POLE
REPORT REVISION

DAYS 1977	****DAILY POLE POSITION X METERS Y METERS	****DAILY SOLUTION**** STANDARD ERROR X METERS Y METERS	****AT-DAILY*** SOLUTION X METERS Y METERS	SATELLITE	NOMINAL UTC-UT1
333. 334.	2.58 1.11	.05 .06	.01-947058.66	1973 81A	244000.00
334. 335.	3.46 .04	.04 .05	.01-953246.66	1970-67A	247000.00
335. 336.	4.21 1.12	.05 .05	.01-947058.66	1973 81A	250000.00
336. 337.	3.00 2.30	.05 .06	.01-953246.60	1970-67A	254000.00
337. 338.	2.72 1.74	.06 .05	.01-947058.66	1973 81A	258000.00
MEAN 336.	3.25 1.75	.05 .05			
STD DEV 336.	.66 .65	.05 .03			
STD ERR 336.	.30 .29	.00 .00			
338. 339.	2.74 1.19	.05 .05	.01-953246.60	1970-67A	262000.00
339. 340.	2.58 1.04	.05 .06	.01-947058.66	1973 81A	266000.00
340. 341.	2.61 1.24	.05 .06	.01-953246.60	1970-67A	270000.00
341. 342.	2.17 1.91	.05 .05	.01-947058.66	1973 81A	274000.00
342. 343.	3.34 1.06	.06 .05	.01-953246.60	1970-67A	278000.00
MEAN 341.	2.69 1.36	.05 .05			
STD DEV 341.	.41 .33	.05 .05			
STD ERR 341.	.18 .15	.00 .00			
343. 344.	1.63 .45	.06 .05	.01-947058.66	1973 81A	279000.00
344. 345.	1.94 .52	.07 .06	.01-953246.60	1970-67A	282000.00
345. 346.	2.06 .46	.05 .06	.01-947058.66	1973 81A	285000.00
346. 347.	2.05 1.33	.05 .05	.01-953246.60	1970-67A	288000.00
347. 348.	1.26 .51	.05 .05	.01-947058.66	1973 81A	290000.00
MEAN 346.	1.82 .66	.06 .05			
STD DEV 346.	.34 .40	.06 .05			
STD ERR 346.	.15 .18	.01 .00			
348. 349.	2.29 1.99	.05 .05	.01-953246.60	1970-67A	294000.00
349. 350.	1.40 .58	.05 .05	.01-947058.66	1973 81A	295000.00
350. 351.	.81 .83	.05 .06	.01-953246.60	1970-67A	297000.00
351. 352.	1.17 .77	.05 .05	.01-947058.66	1973 81A	297000.00
352. 353.	1.52 .49	.05 .05	.01-953246.60	1970-67A	300000.00
MEAN 351.	1.45 .89	.05 .05			
STD DEV 351.	.56 .59	.00 .00			
STD ERR 351.	.25 .27				

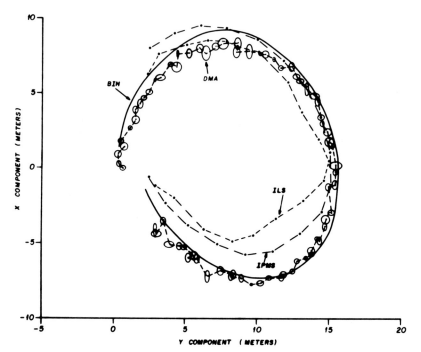

Figure 1. Pole path for 1977.

In the last three lines of each block, only the first three columns are of interest. The first column shows the day number for which the five-day average is being computed. Columns two and three show the weighted averages for x and y, where the weight is taken as $1/\sigma^2$, σ being the two-day standard deviations mentioned above. The line marked STD DEV is the weighted standard deviation of a two-day solution with respect to the five-day mean. The last line, labelled STD ERR, is the previous line divided by the square root of n. It is, therefore, the standard deviation of the five-day mean. Note that the program is presently limited to include only the first four two-day solutions for any given day in the five-day means, even though all available two-day results are listed.

Pole Coordinate Plots

The motion of the pole during the year 1977 may be seen at a glance in Figure 1. This is only a sample. Similar plots for the years 1974 through 1976 will be given in a subsequent report. The Doppler data, now labelled DMA, are easily identified by their one sigma error ellipses. These are the STD ERR of the five-day means shown in Table 3.

Also shown are the polar coordinates from three other sources, namely BIH, ILS and IPMS. They are plotted as solid lines, dashes, and alternating dots and dashes, respectively. As in earlier years, the agreement between BIH and DMA is quite good. However, it must be pointed out that the Doppler data are used, in addition to optical observations, in deriving the BIH pole position results quoted for 1977. The agreement between IPMS and BIH or Doppler is reasonably good. ILS, however, frequently differs from the other three determinations. The discrepancy reaches 3.5 m. The BIH path shown in Figure 1 appears exceedingly smooth. This is due to the fact that only "smoothed' data were available at the time the above plots were prepared.

Figures 2 and 3 permit a comparison of the various polar motion services, separated into the x and y coordinates, for the year 1976. It may be seen that the differences are larger in x than they are in y. Again, plots for the other years will be shown in a later report.

Differences in Pole Coordinates

Anderle (1976b) published plots of differences in the x and y-coordinates of the pole for the time spans 1964 through 1969 and 1970 through 1975. A similar plot, Figure 4, is given in this report for the interval 1974 through 1977, but in x only. This diagram shows the differences in the four pairs BIH-ILS, DMA-ILS, DMA-BIH, and DMA-IPMS quite clearly. By and large, the y-coordinates agree well, except for the ILS excursions in 1976 and 1977. In x, however, all four pairs show significant biases. ILS again shows some large variations with respect to BIH and DMA.

Tables 4 and 5 are a continuation of similar information published by Anderle in earlier reports. They list the yearly average difference for each of the four pairs being compared, as well as the standard deviation of the individual difference with respect to the annual mean. Individual points involving either ILS or IPMS would be 18 days apart, while DMA-BIH is formed every 5 days. Footnotes to Tables 4 and 5 contain additional information concerning data sources and reference frames.

THE CHANDLER PERIOD

It is well known that the principal periodic components of the motion of the pole are the Chandler period and the annual term. In order to determine the former, Anderle (1977, office memo) adapted an existing program to fit to the data an expression of the form

$$X_{comp} = X_o + A_s \sin\left(\frac{2\pi}{365.25}\right)t + A_c \cos\left(\frac{2\pi}{365.25}\right)t$$

$$+ C_s \sin\left(\frac{2\pi}{P_c}\right)t + C_c \cos\left(\frac{2\pi}{P_c}\right)t$$

and a similar equation for y. P_c is the unknown Chandler period, and

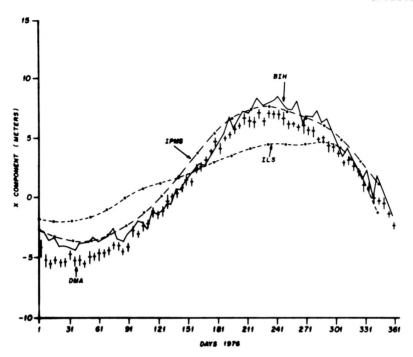

Figure 2. x component of polar motion for 1976.

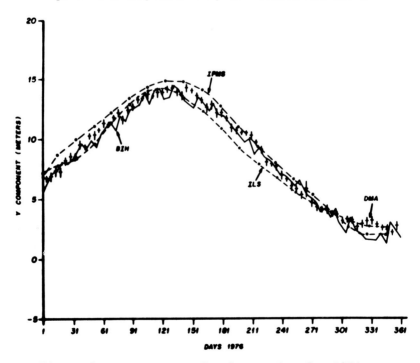

Figure 3. y component of polar motion for 1976.

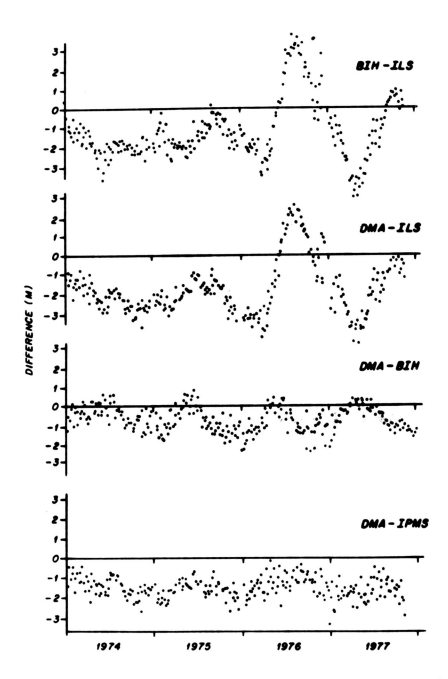

Figure 4. Differences in estimates of the x component of pole position 1974-1977.

Table 4. Average differences in pole position by year.

	MEAN DIFFERENCE (m)						STD DEV OF DIFF (m)					
	X Coordinate			Y Coordinate			X Coordinate			Y Coordinate		
YEAR	DMA -BIH	DMA -ILS	BIH -ILS	DMA -IPMS	DMA -BIH	DMA -ILS	BIH -ILS	DMA -IPMS	DMA -BIH	DMA -ILS	BIH -ILS	DMA -IPMS
1964[1]	1.5	-.5	-1.7	.4	-.0	.3	.8	.2	1.9	1.4	1.1	1.8	1.4	1.5	1.0	1.5
1965[1]	1.6	.7	-.8	.9	.8	.8	.1	.9	.8	1.1	.5	.8	.9	1.6	.8	1.5
1966[1]	.0	-.7	-.6	-.9	.1	.7	.1	.1	.5	1.3	.8	.5	.6	.6	.9	.8
1967[1]	-.5	-.7	-.2	-1.2	-.3	-.2	.1	-.5	1.9	1.8	.9	1.9	1.4	1.9	1.0	1.5
1968	-.7	-.8	-.2	-1.5	.0	-.4	-.3	-.1	1.8	1.1	1.0	1.1	1.5	1.3	.9	1.2
1969	-.3	-1.2	-.8	-1.2	-.1	-.4	-.3	-.1	.9	1.0	1.1	1.0	1.3	1.3	.6	1.3
1970	-.4	-1.1	-.6	-1.4	.3	-.3	-.5	.0	1.0	.8	.8	.8	1.0	1.1	.8	1.4
1971	.1	-.7	-.7	-.8	-.7	-.9	-.2	-.9	.8	1.6	1.5	1.2	.9	1.4	1.3	1.6
1972	-.3	-2.0	-1.4	-1.1	.2	-.2	-.2	-.7	1.0	.9	.9	.7	.7	.9	.8	1.0
1973	-.3	-1.9	-1.5	-.9	-.2	.0	.2	-1.2	.6	1.0	1.2	.8	.5	.5	.6	.6
1974	-.4	-2.2	-1.8	-1.4	-.2	.4	.6	-.6	.5	.7	.7	.5	.5	.8	.7	.4
1975	-.7	-2.1	-1.4	-1.7	.0	.2	.2	-.3	.7	.6	.7	.5	.6	.6	.7	.5
1976[3]	-.9	-.7	.2	-1.3	.3	.5	.2	-.4	.7	2.2	2.3	.6	.5	.9	.9	.8
1977[3,4]	-.6	-2.0	-1.6	-1.7	.1	.7	.6	.3	.6	1.3	1.7	.6	.3	1.4	1.5	.7

ILS Positions are from IPMS annual report and BIH positions are final raw positions from annual report except as noted below.
[1]BIH positions for 1964-1967 are smoothed astronomical positions given in 1969 and 1970 annual reports.
[3]ILS and IPMS positions for 1976 and 1977 are from the Monthly Notes of the International Polar Motion Service (preliminary data).
[4]BIH positions for 1977 are preliminary raw values from Circular D. Note that the positions were computed using DMA data weighted solutions.

Table 5. Coordinate systems and gravity fields.

YEAR	DMA COORDINATE SYSTEM	DMA GRAVITY FIELD	SOURCE OF DMA POLE POSITION
1964	NWL-9D	NWL-9B	NWL TR-2734, 2952 (9 Mean Positions)
1965	NWL-9D	NWL-9E	NWL TR-2734, 2952 (13 Mean Positions)
1966	NWL-9D	NWL-9E	NWL TR-2734, 2952 (6 Mean Positions)
1967	NWL-8D[5]	NWL-8B / NWL-8D (20 Feb)[5]	Preprint[5]
1968	NWL-8F (19 Jan)[6]	NWL-9H (18 Apr)[6]	Preprint[6]
1969	NWL-8F[7,8]	NWL-8H[7,8]	NWL TR-2734[7]
1970	NWL-9C (20 Dec)	NWL-9B (13 Feb)	NWL TR-2734
1971	NWL-9D (18 Oct)	NWL-9B[9]	NWL TR-2734
1972	NWL-9D	NWL-9E	NWL TR-2952
1973	NWL-9D	NWL-10E (2 Jan)	NWL TR-3181
1974	NWL-9D	NWL-10E	DMA Weekly Reports
1975	NWL-9D	NWL-10E	DMA Weekly Reports
1976	NWL-9D	NWL-10E	DMA Weekly Reports
1977	NWL-9Z-2 (15 Jun)	NWL-10E	DMA Weekly Reports

[5]Mean corrections of -2.06 and 1.48 m were added to 1967 NWL-8D X and Y pole positions, respectively, based on comparisons with 12 NWL-9D results given in TR-2734 and 2952.
[6]Mean corrections of -2.37 and 2.04 m were added to 1968 NWL-9D X and Y pole positions, respectively, based on comparisons with 11 NWL-9D results given in TR-2734 and 2952.
[7]Mean corrections of -.07 and -2.35 m were added to 1969 NWL-8D X and Y pole positions, respectively, based on comparisons with 12 NWL 9-D results given in TR-2734 and 2952.
[8]TR-2734 gives pole positions for 1969-1970 computed after adjusting NWL-8F latitude residuals to NWL-10D system.
[9]DMA pole positions for 1964-1966 and after August 1971 are based on simultaneous solution for orbit constants and pole position rather than sequential solutions.

the X_0, A and C are five numerical coefficients to be determined by least-squares fits. One assumes a value for P_c, obtains an expression for X_{comp} and Y_{comp}, and forms the residuals and their rms. This is repeated for several values of P_c, and a parabola is fitted to three such pairs of points. Finally, one computes the value of P_c for which the rms parabola has its minimum. Obviously, to obtain P_c directly from a least-squares solution is more elegant, but the above procedure permitted the use of existing coding.

Figure 5 depicts one of the curves discussed above. The legend shows that 12 years of data were used for the three astronomical sources, while only six years of Doppler data were available for this analysis. It can be seen that the astronomical services agree quite well. The minima are near 432 days, and they are well defined. The Doppler curve in the y-coordinate is of dubious value. It was quickly found that the short six-year time span is responsible. Solutions for the three astronomical sources over six years produced results comparable to the Doppler curve.

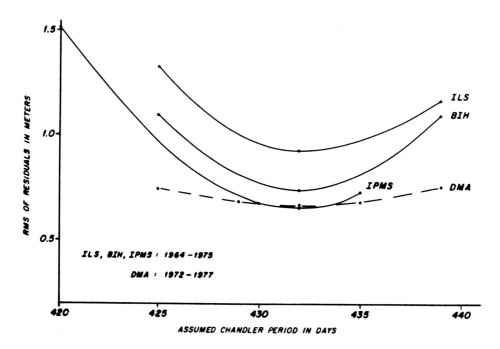

Figure 5. Residuals in y component after 5-parameter fit.

Table 6. Chandler period.

	$P_c(x)$	$P_c(y)$	P_c
ILS	431.71	432.87	432.29
BIH	431.94	432.00	431.97
IPMS	431.77	431.79	431.78

Mean value: $P_c = 432.0 \pm 0.2$ days.

Table 6 contains the results of the P_c computations explained above. In obtaining the averages and the mean value, unit weight was assumed. The error bound of $0^d.2$ was calculated from the scatter of the six individual values.

Although of questionable value, P_c was also computed from the DMA y-curve. It yields $432^d.2$, in reasonable agreement with our adopted values of $432^d.0$.

Our determination is also in good agreement with Markowitz (1976), who obtains $432^d.02 \pm 0.15$ days. It compares reasonably well with Vicente and Currie (1976), who quote 433.2 ± 0.8 days.

The above mentioned residuals are believed to contain other periodicities. Bowman and Leroy (1976), among others, have performed a spectral analysis of the x and y-components themselves, with the following results:

Table 7. Bowman/Leroy spectral analysis.

Frequency (cycles/year)	Period (days)	Amplitude (meters)
0.85	430	5.84 ± 0.6
1.0	365	4.84 ± 0.6
1.3	280	0.49 ± 0.6
2.0	180	0.23 ± 0.6
2.5	145	0.12 ± 0.6
4.0	90	0.11 ± 0.6

Their analysis is based on five years of Doppler data. Considering our earlier difficulties with such a short time span, perhaps considerable strength could be added to the solution by including astronomical data. A particularly attractive time span would be 13 years, corresponding to almost exactly 11 Chandler cycles. However, reliable Doppler data does not yet exist for such an interval.

ADVANTAGES AND DISADVANTAGES OF DOPPLER

Doppler observations are taken day and night, and under any cloud cover. This all-weather capability is one of its major assets. Doppler data are also less sensitive to tropospheric effects than are optical observations. Moreover, they are independent of star catalog position errors. Perhaps Doppler's greatest value lies in the fact that it adds a totally independent pole position determination to the classical methods.

Systematic error due to an inadequate knowledge of the gravity field is the major disadvantage of Doppler. Results are also affected by changes in the station network and atmospheric drag variations during a two-day span. Computing Doppler pole positions is quite expensive. At the present time, however, they are obtained as by-products in orbit improvement runs performed by DMA. Finally, although TRANSIT satellites have shown remarkable endurance, their life time is finite.

FUTURE PLANS

The planning of drag-free satellites is underway at the Applied Physics Laboratory, Johns Hopkins University. Once operational, effects due to drag would be eliminated and a better gravity field could be determined, resulting indirectly in better orbits and pole positions.

The Earth gravity field is continuously being improved, especially by NASA. NSWC has also begun work on a major new geodetic solution. Other improvements in the mathematical model are planned, especially better representations of the various tidal effects.

SUMMARY

Computations of polar coordinates from Doppler observations have been performed in recent years by DMA. During the first half of 1977 as many as five satellites were observed. The standard deviation of a two-day polar coordinate solution is now better than 40 cm, that for the five-day mean under 20 cm. Agreement between the four services ranges from excellent to only fair. There are no significant problems in the y-coordinate, except a 1.5 m standard deviation in 1977 for comparisons involving ILS. The x-coordinate shows both large biases and standard deviations.

It is found that six years of Doppler data are not enough to derive a reliable Chandler period. Hence, twelve years of data from the three astronomical services were taken to compute a Chandler period of 432.0 ± 0.2 days. Residuals suggest the existence of additional periodic terms.

REFERENCES

Anderle, R. J.: 1970, "Polar Motion Determinations by U. S. Navy Doppler Satellite Observations", Naval Weapons Laboratory Technical Report TR-2432, Dahlgren, Virginia.
Anderle, R. J.: 1972, "Pole Position for 1971 Based on Doppler Satellite Observations", Naval Weapons Laboratory Technical Report TR-2734, Dahlgren, Virginia.
Anderle, R. J.: 1973a, Geophysical Surveys 1, p. 147.
Anderle, R. J.: 1973b, "Pole Position for 1972 Based on Doppler Satellite Observations", Naval Weapons Laboratory Report TR-2952, Dahlgren, Virginia.
Anderle, R. J.: 1976a, "Comparison of Doppler and Optical Pole Position over Twelve Years", Naval Surface Weapons Center Technical Report TR-3464, Dahlgren, Virginia.
Anderle, R. J.: 1976b, Bulletin Geodesique 50, p. 377.
Anderle, R. J.: 1976c, "Polar Motion Determined by Doppler Satellite Observations", paper presented at meeting of IAU Commission 19, Grenoble.
Anderle, R. J., and Beuglass, L. K.: 1970, Bulletin Geodesique 96, p. 125.
Beuglass, L. K.: 1974, "Pole Position for 1973 Based on Doppler Satellite Observations", Naval Surface Weapons Center Technical Report TR-3181, Dahlgren, Virginia.
Beuglass, L. K., and Anderle, R. J.: 1972, in S. W. Henriksen, A. Mancini, B. H. Chovitz (eds.), "The Use of Artificial Satellites for Geodesy", American Geophysical Union, Washington, D. C.
Bowman, B. R., and Leroy, C. F.: 1976, in "Satellite Doppler Positioning", Proceedings International Geodetic Symposium, October 1976.
Kershner, R. B.: 1967, in "Practical Space Applications, Advances in the Astronautical Sciences", vol. 21, p. 41.
Markowitz, W.: 1976, "Comparison of ILS, IPMS, and Doppler Polar Motions with Theoretical", Report to IAU Commissions 19 and 31, Grenoble.
Nouel, F. et al.: 1978, "Determination of Polar Motion by Doppler Tracking of Artificial Satellites", CRGS MEDOC Bulletin, January 1978.
Vicente, R. O., and Currie, R. G.: 1976, Geophys. J. Roy. Astron. Soc. 46, p. 67.

DISCUSSION

S. Debarbat: Pouvez-vous donner l'ordre de grandeur du changement dans la précision avec laquelle le pôle est déterminé lorsqu'il y a changement dans l'emplacement d'une station ou changement de satellite observé?
C. Oesterwinter: I have not conducted any experiments to assess the magnitude of these effects. I guess that they would change the coordinates of the pole by only a few centimeters.
J. D. Mulholland: The 1800 terms in the gravitational field correspond approximately to a 40th degree harmonic field. Anderle reported last year that this was necessary, but later retracted. Will

you please comment on this?

C. Oesterwinter: A gravitational field with around 1800 terms is required to compute SEASAT orbits of sufficient accuracy to take full advantage of its 10 cm radar altimeter.

Ya. S. Yatskiv: BIH uses the Doppler observations for determining x, y. IPMS does not. Is this the reason for the better agreement between the BIH and DMA results?

C. Oesterwinter: No. We have used the BIH results based only on optical astrometry.

D. D. McCarthy: I understand that there is a possibility that increasing solar activity may affect the Doppler results; could you comment?

C. Oesterwinter: This possibility is currently being investigated.

IRREGULARITIES OF THE POLAR MOTION

Bernard Guinot
Bureau International de l'Heure
Paris, France

The independent series of coordinates of the pole obtained by the Defense Mapping Agency by observations of Transit satellites (DOP) and by the BIH using classical astrometry only (AST) are considered. They mainly differ by a periodic annual term. After removal of the mean annual difference from AST, the two series are compared. They show common and remarkable irregularities. A derivation of the excitation functions from DOP and AST shows also a good correlation. In the course of this study the constant improvement of DOP, which is now much better than AST, was noticed. The author expresses the opinion that the annual difference between DOP and AST is due to AST, which should therefore be calibrated against DOP.

1. INTRODUCTION

We will consider the irregularities of the polar motion which remain after filtering out the short term noise, up to about three months. The study of the independent series of pole coordinates obtained by classical astrometry and by Doppler observations of artificial satellites will show whether these irregularities are real or due to systematic errors in the observations or in their processing.

It is well known that classical observations of Universal Time and latitude are subject to annual errors. The semi-amplitude of these errors is typically of the order of 0.003s and 0."05, but cam reach 0.010s and 0."10, even for good instruments. This fact has to be taken into account when deriving the coordinates of the pole, x and y, from these observations and when comparing them with the data of other techniques.

2. THE COORDINATES OF THE POLE SUBMITTED TO THE STUDY

The data for astrometry (AST) are the coordinates of the pole of table 6 (until 1972.0) and 6B (after 1972.0) of the BIH Annual Report, where the Doppler data are not used for computing the raw 5-day values of x

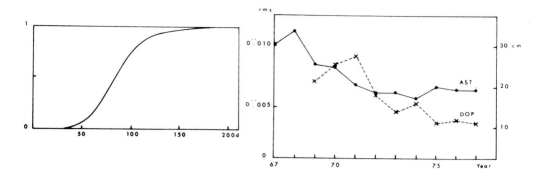

Figure 1. Relative amplitude of a sine wave after filtering, as a function of the period.

Figure 2. Rms differences between filtered and raw values.

and y. However, the Doppler data enter in the determination of the systematic corrections to astrometric measurements of latitude by the following amount:

until 1972.95	0%	for 1975	14%
for 1973	14%	" 1976	13%
" 1974	16%	" 1977	24%.

AST is therefore slightly dependent on Doppler, concerning the mean values of x and y and their annual variations. The 5-day raw values of Tables 6 and 6B are filtered using the Vondrak's method (Vondrak, 1969, 1977). Figure 1 shows the ratio of the amplitude of a sine wave after filtering to the one before filtering, as a function of the period, in the present study. Normal points for x and y are then interpolated for every 0.05 year (normal dates).

The rms differences between filtered and raw values at normal dates are of the order of 0".010 (for x and y) in 1967, then decrease to 0".007 until 1971 and remain fairly constant (Figure 2).

The data for Doppler (DOP) are the coordinates of the pole obtained by the Defense Mapping Agency (USA) from Doppler observations of Transit satellites. The bi-daily solutions are first averaged over 0.05 year intervals, any bias being avoided by the use of a model of the polar motion before averaging. Then a filtering with the Vondrak's method is applied with the same characteristics as for AST. The rms differences between filtered and raw values are shown in Figure 2. The progressive improvement of DOP is remarkable.

3. EXPRESSION OF AST AND DOP IN THE SAME SYSTEM

Table 1 gives the development of DOP - AST in constant, annual and semi-annual terms for each year. The averages of the coefficients are

Table 1. Annual developments of DOP - AST, in 0".001: (t in years)
DOP - AST = a + b sin 2πt + c cos 2πt + d sin 4πt + e cos 4πt.

year	x					y				
	a	b	c	d	e	a	b	c	d	e
1969	- 5	+ 9	- 6	+ 4	-16	+66	- 8	+ 2	+ 2	0
1970	- 7	+25	-22	- 5	+ 4	+14	- 4	+10	+14	+22
1971	- 1	+23	-11	- 3	- 7	-14	-26	-15	-11	+ 3
1972	- 9	+ 3	-21	-12	0	+ 4	+ 6	- 1	+ 5	+ 8
1973	-13	+ 7	- 6	- 5	- 5	- 8	+ 6	- 6	+ 5	+ 4
1974	-14	+ 8	-12	+ 1	+ 7	- 5	+ 9	- 1	- 1	- 2
1975	-24	+18	-22	- 5	+ 8	+ 2	+11	- 1	+ 2	- 4
1976	-27	+14	-25	- 9	+ 6	+10	+ 5	+ 5	- 5	+ 3
1977	-20	+16	-25	0	0	+ 7	+13	+ 4	+ 9	0
1972-77	-17.6	+11.0	-18.4	- 5.0	+ 2.4	+ 1.6	+ 8.4	0.0	+ 2.5	+ 1.7
st. dev.	2.8	2.4	3.2	2.2	2.0	2.9	1.3	1.6	2.0	1.8

computed over 1972-1977, eliminating the first data of DOP which were not homogeneous. The standard deviations agree fairly well with what can be expected from the uncertainties of DOP and AST, and there is little evidence of systematic variations. The expressions of Table 1, where a, b, ..., e are replaced by their mean values, are used as periodic annual corrections to express AST in the system of DOP.

Figure 3 shows the motion of the pole by AST, after correction. This motion is much more irregular from 1972.0 to 1978.0 than from 1967.0 to 1972.0. These irregularities are also found in DOP, as shown by some examples in Figure 4. Local, relatively small, curvatures appear in particular in 1972.60, 1972.85, 1975.00, 1976.35, 1976.65, 1977.25, in both AST and DOP. The time resolution is not sufficient to decide whether these local curvatures are the result of smoothing angular points. This is one of the reasons why more precise methods are needed.

4. EXCITATION FUNCTIONS OF THE POLAR MOTION

It was attempted to derive numerically the modified excitation functions for the motion of the pole (Munk and MacDonald, 1960), separately for DOP and AST. The normal values of x and y were derived using third order differences and the Chandler period was assumed to be 1.19 year. Figure 5 shows these functions, the signs being those used by Munk and MacDonald. They exhibit many common features, especially for ψ_2 which has larger variations than ψ_1. Remarkably large excitation in ψ_2 appeared in 1976/1977.

The modified excitation functions were computed since 1967.0, from AST, but in the system of DOP. Their values can be sent on request. Using

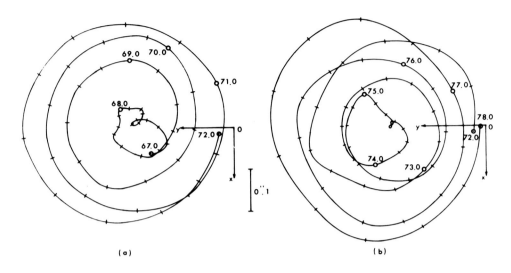

Figure 3. Motion of the pole from astrometry only.

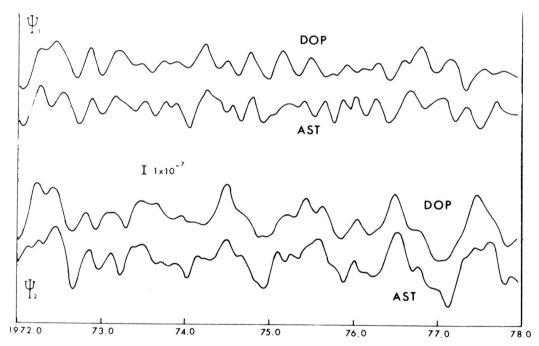

Figure 5. Excitation functions derived separately for Doppler and Astrometry.

as time argument the longitude of the Sun Θ, the mean annual functions develop into (in units of 10^{-8}):

IRREGULARITIES OF THE POLAR MOTION   283

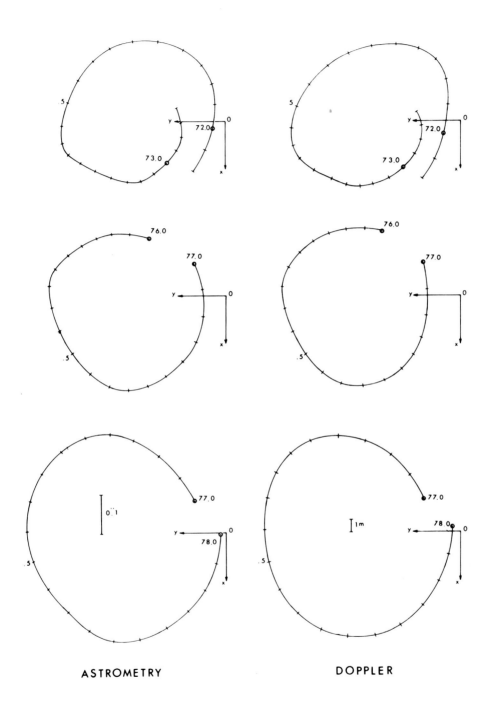

Figure 4. Examples of irregularities of the polar motion.

$$\psi_1 = 5.0 \cos\Theta + 0.1 \sin\Theta + 3.0 \cos2\Theta + 2.2 \sin2\Theta.$$
$$\psi_2 = 2.6 \cos\Theta + 13.1 \sin\Theta - 2.8 \cos2\Theta - 1.0 \sin2\Theta.$$

4. CONCLUSIONS

Concerning the geophysical aspects, we conclude that large and short-lived (less than a month) disturbances of the polar motion occur from time to time. The study of their fine structure justifies the efforts for obtaining more precise measurements. It was shown also that reliable values of the excitation functions of the motion of the pole can be found experimentally from the coordinates of the pole.

Concerning the operation of services dealing with the coordinates of the pole, we make the following remarks. The agreement of Doppler with astrometry is good, taking into account the specific annual errors of the latter. If the aim is to provide users with the best coordinates of the pole (and not to study some particular problems attached to each method), it is permissible to mix the data of these two sources, with the relative weights they deserve. The Doppler results are much more precise nowadays than the results derived from the whole set of 80 astrometric instruments; a fair weighting would give weights 3 to 4 to Doppler and 1 to astrometry. We also believe that the accuracy of Doppler results is better for the annual terms, and that it is at least as good as astrometry for the progressive variations. This opinion is based on the long-term consistency in positions of sites determined from Doppler observations (Anderle, 1975) and on the comparison with astrometry.

Therefore, astrometric measurements of the pole appear mainly as a safeguard against possible interruption of the Doppler data. It would be logical to calibrate the astrometric annual terms against Doppler. At the BIH, it was safer to wait until other techniques considered confirm the accuracy of Doppler results before doing so. But on account of the delays in the implementation of new techniques and the constant improvement of the Doppler results, the question arises whether the astrometric system should be immediately converted into the Doppler system.

REFERENCES

Anderle, R. J.: 1975, Naval Surface Weapon Center, Dahlgren Lab., Report TR-3433.
Munk, W. H. and MacDonald, G. J. F.: 1960, "The Rotation of the Earth", Cambridge Univ. Press, p. 41.
Vondrak, J.: 1969, Bull. Astron. Inst. Czech. 20, p. 249.
Vondrak, J.: 1977, Bull. Astron. Inst. Czech. 28, p. 84.

DISCUSSION

Ya. S. Yatskiv: When calculating the excitation function, did you use the complex Chandler frequency?

B. Guinot: No. I used the real value of the Chandler frequency.

S.K. Runcorn: It is important for the geophysicist to know what is the minimum disturbance, in amplitude and time scale, which would be definitely detected by present techniques.

B. Guinot: I have not attempted to find the minimum disturbance which is observable, because it depends on the duration of the disturbance.

A.R. Robbins: You said that if Doppler satellites would continue operating for ever then astronomical observations could be discontinued. How about maintaining the reference frame?

B. Guinot: The classical observations do not seem better than Doppler for maintaining the reference frame. They are affected by changes in the location of instruments, and by changes of programs. Even in the case of the ILS organization, a spurious drift of the pole is suspected by many authors. The spurious drift of the Doppler pole does not appear worse.

F.P. Fedorov: Can you say something about the secular term, or is the interval of concurrent classical and Doppler observations too short?

B. Guinot: Precise comparisons have been made only from 1972; the interval is too short.

REASONS AND POSSIBILITIES FOR AN EXTENDED USE OF THE TRANSIT SYSTEM

P. Pâquet and C. Devis
Observatoire Royal   Université Catholique
de Belgique   de Louvain

SUMMARY

Observations of Earth rotation and polar motion are now performed on a routine basis by classical astronomical methods and by the satellite Doppler technique.

In the near future the laser is likely to realize its old objective of determining ER and PM with an accuracy of a few milliseconds of arc through ranging to Lageos. VLBI is still in development.

To permit investigation of possible systematic differences, all these techniques should be used simultaneously for a period whose length should not be set in advance.

The Doppler method can still be improved and could be easily introduced at many observatories in support of the Medoc campaign. We present here two possible methods of data condensation which, if implemented, would permit much of the data analysis to be done at the tracking stations and would reduce to a very few the number of parameters required by the central computing agency.

1. NON A PRIORI SELECTION OF A SPECIFIC METHOD OF OBSERVATION

During the UGGI General Assembly of Grenoble (1975) Prof. P. Melchior, considering the availability of satellite techniques and their good results, initiated a free discussion centered on the use of these new methods and their impact on the future evolution of International Services devoted to Polar Motion (PM) and Earth Rotation (ER) studies. At the next IAU General Assembly Prof. Robbins (1976) gave a review of the various proposals given by many geodesists and astronomers. The main conclusions were that it was at present impossible:

- to answer the question "which techniques must be maintained or selected?"

- to define the length of the period during which the coexistence of classical (PZT, astrolabe..) and new instruments (Doppler, laser, interferometer..) should be maintained.

Since that time the reasons and needs for more precise measurement of the Earth's angular position have been often considered, and summarized by P.Melchoir (1976), J. Popelar (1976), A.C. Schulteis (1977), P. Paquet (1978). From many discussions it comes out that a 5 cm precision in the pole positioning and 0$\overset{s}{.}$0001 for the Earth rotation velocity are requested over time intervals extended from few hours to one day. These constraints are the proposed objectives to be reached within the next 10 years. Precise polar coordinates are of immediate use for space research and geodesy but not necessarily on a continuous basis. However, in the longer term, for humanitarian and economic reasons, programmes considering earthquake predictions (plate deformation control) and climatic variations could require continuous knowledge of the parameters fixing the observers' positions.

It is important to know if a high accuracy is continuously necessary or if very sophisticated techniques will be requested only for particular circumstances. In the meantime less accurate methods could perhaps monitor the Earth rotation and polar motion.

The fascination of space research and the possibilities for increasing the accuracy with which ER and PM could be measured have produced a competition between scientists, each defending his preferred technology: astronomical, Doppler, Laser or interferometry. In practice all these techniques are complementary and may be considered as developments in the evolution of the best system (?) of observation whose use would be just to check if Earth motions are those to be expected from theory.

What is the present state of the techniques for Earth rotation and polar motion measurements?

For several years an accuracy of $0\overset{''}{.}01$ has been obtained by classical astronomy and, as improvements are in progress, an accuracy between one hundredth and few thousandths of a second of arc can be expected during the next decade. What is desirable is to reduce progressively the observer functions by installation of more automated instruments like the PZT and the Chinese astrolabe.

For the Laser, confronted by the difficulties of ranging to the Moon on a continuous basis, a new solution is now possible with ranging to the Lageos satellite. To check the expected performance on Lageos, a request for a several years campaign has been expressed at the Workshop on Space Oceanography, Navigation and Geodynamics (SONG, 1978) organised by the European Space Agency.

Long baseline radio interferometry (VLBI) is in a similar

situation to that of the Moon laser at the beginning of the 1970's A few experiments have been realized, some capabilities have been demonstrated and during the next years it remains to improve the technology (Schilizzi, 1978). Several years ago short baseline interferometry (SBI) was able to measure the angular Earth position with similar precision to that obtained with a PZT (Ryle and Elsmore, 1973). Besides these systems the radio Doppler method is well established and is the only new technique producing pole coordinates on a routine basis. The Doppler low cost instrumentation is already used in many disciplines such as astronomy, geodesy, navigation, and geology, and the Workshop SONG (1978) requested experiments using the Doppler method with a third frequency near 2 GHz.

Considering the state of art of these methods, no selection of a specific technique can be made at present and we must consider that their coexistence is fortunate. It is an excellent arrangement for detecting systematic deviations. From the same point of view it is not realistic to predetermine an overlapping period during which the different methods are to be used. This overlapping must be long enough to compare the results and reach conclusions we may trust. However, considering the urgent request for higher precision in determination of the Earth's position, we think that Commission 19 should encourage and recommend the use of new techniques during the next few years.

Having this in mind we will propose here some methods for which trials are in process for a possible extension of the Doppler system. By extension we mean to allow for many observatories to be easily introduced in a tracking network, to perform their own preprocessing and so reduce the amount of data transmission and the task of the Central Computing Agency. This is indeed the usual procedure in classical astronomy: each station reduces its own set of observations and only the final results are transmitted to BIH and IPMS.

2. REASON FOR AN EXTENDED USE OF THE DOPPLER METHOD

If the technical performance of the Transit satellites remains in the present state is it possible to improve the Doppler results? There are several reasons for believing that the theory and modelling of perturbations acting on the satellite motion and signal propagation could still be improved:

- B. Bowman and C. Leroy (1976) estimated at 40 cm the possible error on pole positions deduced from their analysis. These authors also showed that occasionally separate solutions deduced from different Transit satellites can give pole positions differing by between 50 cm and 1 m. This inhomogeneity was not explained, but they attribute it to resonance effects, and to a different number and distribution of stations for the data collected on each satellite.

- if external errors are here understood as those deduced from comparisons of several similar experiments, and not from results obtained by two different methods, then for Doppler results the ratio between external and internal errors was estimated by Paquet (1973, 1977) to be about 2.5.

- the technology of tracking stations could be improved.

- the present results obtained by DMA are based on the observations performed by 17 TRANET stations acquiring data from only two of the five available Transit satellites. A lot of data which are not used, and generally not even collected, could be included in the computations.

- GRGS has restricted the MEDOC experiment to only one Transit satellite.

All this demonstrates that the models used for the data analysis could be improved with the support of a great amount of data which remains available. However, to use all the data acquired at each station, the central computing center has to solve two main difficulties:

- To produce results without unreasonable delay the data collection must be performed by telex or some other fast way.

For a tranet station the number of useful bits collected during each pass is about 2 000. For JMR or Marconi receivers one pass is composed of about 10 000 bits of Doppler data. The transmission of all the bits is a first important limitation in cost and volume.

- the work of data preprocessing is very important.

To support Doppler programmes it should be necessary to reduce to a minimum of parameters the data acquired during one pass or a set of passes. These condensed data will have to be deduced by pass preprocessing performed at each station, on the basis of a unified theoretical model of reduction. This unified model must contain:

- the modelling of perturbations to be immediately removed from observations;

- the gravity model used to integrate the approximate orbit, whose initial conditions are to be regularly renewed by the central computing center;

The Computer programmes must be delivered by the same agency.

Advantages of such a solution are that it

- considerably reduces the quantity of data to be transmitted;

- shares the preprocessing tasks between observers;

- increases the number of satellites that can be included in the solution for Earth rotation and polar motion determinations.

Moreover, as suggested by M.Lefebvre and F.Nouel (CNES), if the pass can be reduced to a few parameters, the transmission facilities of the ARGOS satellite will be available; this is being developed as a joint Franco-American (CNES, NOAA, NASA) project for a location and data-collection system. Information is available from the CNES center at Toulouse.

3. POSSIBLE SOLUTIONS

Observatories generally have good computing facilities, which would allow calculation of a satellite orbit with the best available methods of numerical integration and using one extended gravity model taking account of solar pressure, atmospheric drag, Earth tidal effects, Sun and Moon attraction ... The availability of the best predicted orbit makes it possible to perform efficient preprocessing at each station on a pass-by-pass basis. This preprocessing consists of the comparison of the observed quantities with the theoretical ones as deduced from the predicted orbit.

In the Doppler system the data selection generally results from a least squares solution, fitting the pass observations by adjustment of frequency offset ΔF, tropospheric scaling factor, range (R_o, R_2) and along track (L_o, L_2) parameters as defined by Guier (1963) in a reference system with its origin at the satellite position at the time of closest approach (TCA). The axes are:

- R, the range axis from station to satellite

- L, the along-track axis positively oriented with the satellite motion

- Z, the third axis perpendicular to the plane (R, L).

The along-track and range unknowns are called the navigation errors related to the errors on the station and the satellite coordinates. The equations of observations fitting the Doppler residuals ℓ_j are of the form:

(1) $\quad \ell_j = \Delta F + L_o u_{s,o} + L_2 u_{s,2} + \ldots + R_o u_{a,o} + R_2 u_{a,2} + \ldots$

where the coefficients $\underline{u}$ are given in Guier (1963).

The values of these parameters associated with

- the errors, on each parameter, resulting from the solution of the normal equations,

- the number of accepted data after filtering,

- the balance between data acquired before and after TCA,

were used to define the constraints of pass rejection during the European Geodetic Campaigns EDOC-1 and EDOC-2 (Pâquet 1977). With appropriate upper limits they define a set of very selective filters, efficient enough to eliminate all doubtful passes and allowing the surviving passes to be introduced definitively in the matrix developed to determine station coordinates. Anderle (1973) was using similar parameters.

For data condensation this preprocessing is the first necessary task to be done in situ. Two methods of condensation can be employed:-

a) Condensation of a set of passes

After the preprocessing the usual procedure to improve the satellite orbit is to fit the accepted data in a least squares solution which includes as unknowns the 6 constants of orbital integration, the two coordinates of the pole, one drag scaling factor and some complementary station parameters. In the Doppler system, for each station and for each pass, these complementary unknowns are the frequency offset and a tropospheric refraction scaling factor. The final system of normal equations results from adding the individual systems deduced from each accepted pass of the tracking network.

Having a large experience in such an analysis R. Anderle (1973) wrote: "The least squares solution is not iterated under normal circumstances since the prediction errors from the preceding orbital fit rarely exceed 100 m during the 48 hours prediction interval". At present, at the end of the same interval of time, the prediction error seems to be lower than 50 meters; this strengthens Anderle's remark.

From these remarks it becomes clear that a first method of condensation is for each station to perform its own preprocessing and compute, for orbit improvement, its own set of normal equations including all accepted passes during the agreed periods. Even this simplest solution yields a substantial reduction in the number of unknowns. Indeed, if the preprocessing is performed with the best predicted orbit the instrumental and local parameters may be determined by each station.

With the number of unknowns thus reduced to 9, and with one orbit improvement every two days, 54 parameters have to be transmitted to the Central Computing Agency every two days. This corresponds to a maximum of 3000 bits, less than two satellite passes recorded by Tranet equipments.

With this procedure the task of the CCA would be to combine the matrices received from all the stations, solve the system, and return the solution to the observers for the next prediction.

b) <u>Pass by pass condensation</u>

During the preprocessing step the along-track and range parameters determined by separate pass analysis reflect station-position errors and also errors of the predicted satellite orbit. The relations between station-satellite errors and fitted parameters were derived by Guier (1963). The quantities (L, R) of equation (1) are given as linear functions of the station displacement and satellite position errors and their derivatives, evaluated at TCA and expressed in a reference system whose origin is the satellite position at TCA.

These relations have been extensively used with success in geodesy for two main purposes:

- to improve the coordinates of ground stations. The orbit is assumed fixed and the total position errors are attributed to the erroneous initial station position. The European Doppler Campaigns, EDOC-1 and EDOC-2 were treated by this procedure (Pâquet, 1976, 1977) and the results are of the same quality as those obtained by more classical methods.

- W.H. Guier and R.R. Newton (1965) and S.M. Yionoulis et al. (1972) used the relations to improve the Earth gravity field. For this analysis Guier and Newton developed a linear perturbation theory allowing computation of (L, R) in terms of changes of the gravity harmonics, the orbit parameters and the station positions.

The success of these relations suggests their experimental use as a second possibility for data condensation. The linear approximation must be limited to variation of the orbital elements, coordinates of the pole position and Earth rotation velocity, while the gravity field and the station coordinates are fixed. This process, if correctly conducted, allows reduction of the data to be transmitted to less than 300 bits per pass.

REFERENCES

Anderle, R.J.: 1973, *Geophys*. Survey 1, 147.

Bowman, B. and Leroy, C.: 1976, in *Satellite Doppler Positioning*, Proc. Inter. Geod. Symp., U.S. Defense Mapping Agency (DMA) and National Oceanic Survey, NOAA.

Guier, W.H.: 1963, *Studies on Doppler residuals*, Appl. Phys. Lab. TG-503, Johns Hopkins Univ., Silver Spring, Md.

Guier, W.H. and Newton, R.R.: 1965, *J.G.R. 70*, N° 18, 4613.

Guinot B., and Nouel, F.: 1976, in *Satellite Doppler Positioning*, Proc. Inter. Geod. Symp., U.S. DMA and NOAA.

Melchoir, P.: 1976, in *Satellite Doppler Positioning*, Proc. Inter. Geod. Symp., U.S. DMA and NOAA.

Nouel, F.: 1976, *MEDOC experiment*, GRGS scientific report, Toulouse.

Paquet, P., Dejaiffe, R.: 1973, Proc. of Symp. on Earth's Gravitational Field and Secular Variations Position, p. 347, Ed. MATHER and ANGUS-LEPPAN, School of Surveying, Sydney.

Paquet P.: 1976, in *Satellite Doppler Positioning*, Proc. Int. Geod. Symp., U.S. DMA and NOAA.

Paquet, P.: 1977, *Preliminary report on EDOC-2 data analysis performed at the Royal Observatory of BELGIUM*, presented at Journ, Lux. de Geod., Walferdange.

Paquet, P.: 1978, in *Space Oceanography Navigation and Geodynamics*, Proc. of a European Workshop SONG, p. 181, Ed. by S. Hieber and T. Guyenne, ESA.

Popelar, J.: 1976, in *Satellite Doppler Positioning*, Proc. Int. Geod. Symp., U.S. DMA and NOAA.

Robbins, A.R.: 1976, *A future International Earth Rotation Service*, presented at the IAU General Assembly.

Ryle, M. and Elsmore, B.: 1973, Month. Not. R.A.S., 164, p.223.

Schilizzi, R.T. and Campbell D.: 1978, in *Space Oceanography Navigation and Geodynmaics*, Proc. of a European Workshop SONG, p. 329, Ed. by S. Hieber and T. Guyenne, ESA.

Schulteis, A., Sullivan R.J. and Harris R.L..: 1977, *Estimates of benefits and costs of measuring polar motion and Universal Time using VLBI and satellite laser ranging*, Rpt 289, System Planning Corporation, Arlington, USA.

Yionoulis S.M., Heuring F.T. and Guier W.H.: 1972, *J.G.R.* 77, N° 20, 3671.

VERY FIRST RESULTS OF THE MEDOC EXPERIMENT

F. Nouel and D. Gambis
Groupe de Recherches de Géodésie Spatiale, Toulouse.

It is already well known that the coordinates of the pole can be derived not only by the classical astronomical methods but also from analysis of the orbits of artificial Earth satellites. The MEDOC experiment (Motion of the Earth by Doppler Observation Campaign) has been initiated by the Groupe de Recherches de Géodésie Spatiale (GRGS) to provide observations and undertake independent analysis for this purpose. Several organisations are participating.

The very first results, presented here, are based on Doppler observations of a Transit satellite, made at 11 stations around the world during a period of 9 months.

Details of the processing and analysis of the data are also given.

1. OBJECTIVES OF THE EXPERIMENT

"Classical" astronomical methods have been used for many years in the determination of polar motion. The tracking of artificial Earth satellites now provides similar data; the analysis is undertaken by the Defense Mapping Agency Topographic Center (DMATC). It appeared to be desirable to investigate the sensitivity of the results to variations in the parameters used in the dynamical analysis, and the GRGS offered to undertake a similar determination with the following final objectives:

- to determine the extent of, and the reasons for, systematic differences between the results of various computer programs;

- to promote satellite techniques;

- to lay the groundwork for a scientific service.

Doppler data have been collected since January 1977 and the investigations are in progress.

2. THE ORGANISATION

One advantage of the Doppler satellite-tracking technique is that it permits operation in all weathers; moreover, the Transit satellite family and the corresponding Tranet network already exist and provide an operational service. The stations participating in MEDOC were chosen because they were foreign scientific institutions which could contribute not only by sending data but also by their scientific activity.

SMITHFIELD	. Division of National Mapping - Australia
MIZUSAWA	. International Latitude Observatory - Japan
UCCLE	. Observatoire Royal de Belgique - Belgium
OTTAWA CALGARY	. Earth Physics Branch - Canada
HERDON	. DMA Tranet Station - Virginia U.S.A.
UKIAH	. National Ocean Survey - California U.S.A.
MELVILLE	. Shell Canada - Canada (for a short period)

All the data from these sites are collected by DMATC and sent every week to GRGS by DMATC.

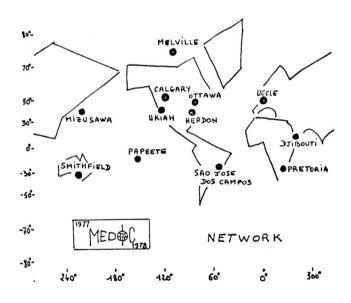

In order to optimize the balance of the network, GRGS (The Institut Geographique National, IGN and Centre National d'Etudes Spatiales, CNES) provides four receivers which are located in:

DJIBOUTI	. Observatoire sismologique - Djibouti Republic
PRETORIA	. French Tracking station NTIR/CNES - South Africa
SAN JOSE DOS CAMPOS	. Instituto de Pesquisas Espaciais - Brasil
TAHITI	. Laboratoire de Geophysique de Pamatai - French Polynesia

These receivers are of the JMR type; they work in automatic mode and the data are recorded on minicassettes which are read at the computing center. At these four last sites "housekeeping" takes about half an hour per day.

3. DATA PROCESSING

3.1. Preprocessing

The raw data received at the computing center come from different types of receiver and must be preprocessed in order to:

- make them compatible with the measured quantity used in the programs; for each Doppler quantity we must have the number of accumulated cycles, the date at which the count starts and the time for which it lasts;

- ensure that all required quantities, such as meteorological information, and linkage with an accuracy of about 50 µs to a coordinated time scale, are known for each block of data;

- eliminate spurious data and measurements taken at low angles of elevation; the entire pass may be rejected, or only a few points near the start and finish.

3.2. Computations

Schematically, computing the polar coordinates for a two-day interval consists in integrating the equations of the motion of the satellite over that time, and in adjusting some of the parameters in these equations to better fit the observed measurements. The reference system is the inertial frame of dynamics. The geographical coordinates of the receivers located on the Earth have to be rotated towards this inertial frame, and in doing this the polar motion must be taken into account. We can either use given values of the components (for example, values from BIH publications) or introduce them as unknowns and try to determine them.

The motion of the satellite is computed from a force model which includes gravitation, due mainly to the Earth but also to the Sun and the Moon, atmospheric drag, solar radiation pressure, and the perturbation of the Earth potential due to the Earth tides. In addition, since we actually use the instantaneous reference system as our reference frame because this makes the calculations easier and saves computing time, we also have to take into account the apparent (Coriolis) forces due to precession and nutation.

The parameters which are determined by the least-squares fit are:

- six quantities giving the shape and orientation of the orbit;

- one scale factor for the atmospheric drag and one for the solar radiation pressure, because the knowledge of the effective area-to-mass ratio of the satellite is poor;

- one frequency offset for each pass over a station, to account for local oscillator variations; and

- two components of the position of the pole.

3.3. Station reference system

The geographical system is defined by the set of the coordinates of the participating stations; but, because these coordinates are also computed from Doppler measurements, the network is strongly correlated, at least at the tens of meters level, with the model of the Earth potential.

So far, in the MEDOC experiment, we have used the GEM X model from the Goddard Space Flight Center, and station coordinates are computed with respect to that model by an iterative procedure in which the orbit and the set of station coordinates are computed alternately. Iteration is stopped when successive positions differ by only a few meters. Variations of the reference system with time show similar internal coherence.

4. POLE COORDINATES

4.1. Results

The coordinates of the pole at intervals of two days have been computed by this method for the period from 1977 January to October. The results are compared in the figure with the BIH smoothed values and the DMATC results; X and Y have their usual meanings, X being positive towards Greenwich and Y perpendicular towards the West in the plane perpendicular to the CIO Z-axis.

4.2. General remarks

The bias between the MEDOC solution and the others is attributed to the set of station coordinates used as the reference system; it will be removed when we have accumulated positions for several years.

The most important jumps in both components are strongly correlated with the failure of several stations. The stability should be improved by the addition of suitably located stations to the network; it will also improve when the set of station coordinates has been refined to the one meter level and is less sensitive to the presence or absence of a single station.

Some preliminary spectral analysis shows significant peaks which correspond to periods having resonances with particular spherical harmonics in the development of the Earth's gravity field; examples are (13,13), (14,13) and (26,27).

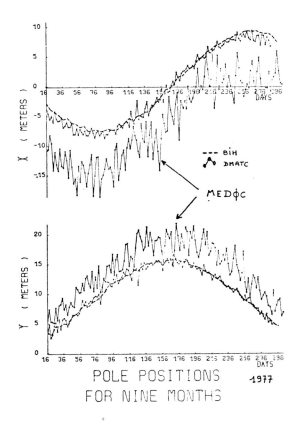

POLE POSITIONS FOR NINE MONTHS 1977

5. CONCLUSIONS AND FUTURE DEVELOPMENT OF MEDOC

The main difficulties arise from the need to maintain the network in routine operation. Receivers must be operated at remote sites at which it is difficult to get them repaired, and non-technical problems often increase the delay before the equipment is returned to use. An increase in the number of receivers will reduce these problems.

The very first results are promising, but the objectives of MEDOC have not yet been reached. Observations are being accumulated and will be used, in the coming years, to construct a model of the Earth's potential which is appropriate for use with the polar satellite used in MEDOC. Analyses will then be made to investigate the influence of each assumed parameter on the derived coordinates of the pole. In addition, experience gained in collecting, transmitting and processing the data will be useful in studies relevant to the establishment of a permanent polar-motion service based on satellite techniques.

References

Anderle, R.J.; Polar Motion determined by Doppler Satellite observations, *Bulletin Géodésique 1976*.

Bowman, B.R., Leroy C.F.; DMATC Doppler Determination of Polar Motion. *International Geodetic Symposium on Satellite Doppler Positioning*, Las Cruces, October 76.

Guinot, B., Nouel, F.; MEDOC Experiment on the French Polar Motion Project. *International Geodetic Symposium on Satellite Doppler Positioning*, Las Cruces, October 76.

DISCUSSION

D.D. McCarthy: How many stations would be required to provide a permanent service delivering an accuracy of 0.01 arcsec, assuming that their distribution was good?

F. Nouel: I suppose that something like 15 stations would be needed if we could start with a good model of the Earth potential; but construction of the model would need more than 15 stations, or another technique such as altimetry.

PART VIII : GEOPHYSICS

THE GEOPHYSICAL INTERPRETATION OF CHANGES IN THE LENGTH OF THE DAY
AND POLAR MOTION

S.K.Runcorn, School of Physics, The University,
Newcastle upon Tyne, NE1 7RU, U.K.

ABSTRACT. The data on the irregular fluctuations in the length of the day and the motion of the pole is of great significance in the geophysicist's task of constructing a model of the earth's interior.

Short term instabilities in the dynamo generating the earth's magnetic field produce what is observed at the surface as the secular variation. These changes induce currents in the lower mantle and the resulting torques appear to be the cause of the irregular fluctuations in the length of the day, although some quantitative problems remain.

The excitation of the Chandler wobble could result from impulsive torques applied to the mantle by very short period (a year or less) local magnetic field disturbances coming to the surface of the core. The alternative mechanism by earthquakes has been much investigated and the possibility of this is still obscure. A test however is available in the polar motion data: a disturbance in its path displaces the subsequent centre of its Chandler motion in the latter theory, but only its amplitude on the former theory. The behaviour of the pole around 1968 supports the core theory, but much more analysis of the polar motion is required.

1. THE LENGTH OF THE DAY.

Spencer Jones (1939) demonstrated from the observed discrepancies in the sun, moon and inner planets that the irregular fluctuations in the length of the day have a time scale of tens to hundreds of years. This is only paralleled in geophysics by that of the geomagnetic secular variation. At that time, not enough was known of the generation of the geomagnetic field and its secular variation for any understanding of the mechanism of these enigmatic changes in the earth's rotation. Their origin had been very puzzling since de Sitter (1927) clearly demonstrated that they could not occur through crustal processes: to change the moment of inertia by the amount to increase the length of the day by 3 m sec, observed just before 1900, would require the raising of another Himalayas.

The discovery by Vestine et al (1947) - or rediscovery as the phenomenon had been known to Halley (1692) and to Bauer (1895) - of the westward drift of the geomagnetic field (about $1/5°$ per year) provided the important clue. Alfven's fruitful idea that, on the cosmic scale, lines of force move with a conducting fluid provides a simple interpretation: the earth's core is at present rotating more slowly than the mantle, about 1 cm/s at the interface, a speed similar to that of the convection in the liquid iron core needed to generate the geomagnetic field by the dynamo process. Clearly a change in the rotation of the core could explain the irregular fluctuations of the length of the day, for assuming conservation of angular momentum of the earth, Runcorn (1955) showed that the change in the length of the day (dT) in sec and the change of the westward drift ($d\omega$) expressed in °/yr are related by

$$dT = 0.067/d\omega$$

The largest change in the length of the day referred to above could be brought about by a 20% change in the westward drift of the core.

Different ways of measuring the westward drift give roughly the same values, except for the rotation of the equatorial dipole which is much smaller and is not really understood. Thus Vestine (1953) was able to trace the variation of the westward drift back to 1820 by determining, from the spherical harmonic analyses of the field from Gauss onwards, the longitudes of the off-centre dipole commonly used to represent the quadrupole term in the field. The latest comparison of the length of the day and the westward drift (Kahle et al 1969) shows a good correlation, supporting the theory that the irregular changes of the length of the day arise from transfers of angular momentum between the core and the mantle. It is not surprising that the exact theory of the nature of the coupling mechanism is still controversial, except that viscous coupling fails by many orders of magnitude.

Recently a new and most interesting correlation has been found by Cazenave & Lambeck (1976) between the irregular changes in the length of the day and the angular momentum of the atmosphere, as determined from surface pressure observations. This has planted doubt in some minds as to the correctness of the above theory, but quantitative considerations suggest that it is unlikely that the changes in atmospheric circulation could be responsible. Further, rather sudden changes in the rotation of the earth's mantle would alter the atmospheric angular momentum as viewed from the earth and would appear to be a satisfactory explanation of the correlation found. The sign predicted by these two alternate theories are opposite and should supply a decisive test: an increase in the westward flow in the atmosphere should be associated with a decrease in the length of the day on the hypothesis that the atmosphere is responsible and with an increase in the length of the day if the core is responsible.

2. CORE MANTLE COUPLING

Two forms of core-mantle coupling have been discussed: electromagnetic couples produced by induced electric currents in the lower mantle, which is a semiconductor, and hydrodynamic coupling produced in the core by hypothetical undulations on the core mantle boundary of some 10 km high and 100-1000 km in wavelength. The latter suggestion, due to Hide (1969), arises from an attempt to explain a correlation between the geomagnetic non-dipole field potential and the geoid, both cut off above the 4th degree, and the geoid displaced through $160°$. The electromagnetic coupling encounters some quantitative problems as pointed out by Roden (1963), Rochester (1960) and Roberts (1972), but in extrapolating the known surface field to the core boundary, the strength of the varying fields in the lower mantle produced by changing eddies in the core may be underestimated. Runcorn (1970) has pointed out that rapidly changing fields may arise from the core, as Alfven wave velocities of magnetodynamic disturbances (about 100 m/sec) divided into the length scale of core eddies (100 km) yield time scales of a month. The lower mantle below 1000-2000 km, which is likely to have conductivities of $100 \Omega^{-1} m^{-1}$, would screen such rapidly changing fields and they would be undetected at the Earth's surface.

In understanding the theory, much depends on the analysis of the observations. De Sitter (1927) represented the discrepancy between the observed and theoretical longitude of the Moon (where time was measured in subdivisions of the day) as a series of straight lines, Brouwer (1952) by a series of parabolic arcs: the former require impulsive torques to cause discontinuities in the Earth's rotation and the latter abrupt changes in the magnitude of torques, neither being physically plausible. Smoothing techniques inevitably remove any sharp changes, even if they are present and a new method of analysis to determine the time scale over which changes take place is needed. The geomagnetic secular change is a regional phenomenon, the areas of rapid change establishing themselves in a few tens of years, last for a few hundred years drifting westwards, and disappear to be replaced by other isoporic centres - Elsasser's analogy to meteorological charts is apt. Interpretation by a number of eddies each generating a growing or decreasing dipole field in the outermost core has been made. Supposing a dipole growing, induced current loops in the mantle would develop in a time equal to σa^2, where σ is the lower mantle conductivity and a the size of the eddy, or about 1 month - 1 year, and then would remain constant until the dipole growth ceased. This would yield a constant torque over the lifetime, say some tens of years, of the eddy. If, however, a constant field surfaced in the core an impulsive torque would be produced with time scale of about 1 year. Examination of the length of day variations suggests that both are present.

3. THE CHANDLER WOBBLE

The course of the impulses which generate the Chandler wobble is puzzling and earthquakes and the atmosphere have been suggested. Yet if electromagnetic torques must be invoked to explain the changes in the length of the day and if some are impulses, then, bearing in mind the complexity of the secular variation, it is not reasonable to suppose the electromagnetic torques on the mantle are axial torques. Components in the equatorial plane will generate the Chandler wobble. As Runcorn (1970) pointed out, there is a considerable difference in the physics between constant and impulsive torques: the former generate a forced nutation and as Mintz & Munk (1951) showed these must be negligible contributions to the polar motion, but the latter, supposed of the same order as the impulsive torques along the earth's axis required to explain the sharper changes in the length of the day, e.g. that just prior to 1900, would be effective in exciting the Chandler wobble.

Again the analysis of the data is the key to a decision between the different mechanisms, as Runcorn (1969) showed. To generate the Chandler wobble, consider the axis of figure F, the axis of instantaneous rotation I and the axis of angular momentum M to be at first coincident. If an earthquake excites the Chandler wobble, F is suddenly displaced and I and M, which are very close together rotate in a circle around a new centre F, see Fig. 1(ii). If an impulsive torque is applied in the equatorial plane to the mantle, F remains fixed and M (and I) are suddenly displaced and then move in a circle around the same centre. This is shown in Fig. 1(i). What is needed is an analysis of the Chandler component of the polar motion to examine whether disturbances in the path are of the former or latter type. Guinot (1972) using a method to remove the 12 monthly term in the polar motion due to atmospheric excitation finds a case in 1968 when the polar path sharply diverges and then continues in a circular path round the same centre. This is clearly explicable only in terms of an impulsive torque (with a time scale of at most a few months), and is not what would be expected from an earthquake. It seems unlikely that the atmosphere could provide such an impulsive torque.

Another phenomenon in polar motion data of interest to the geophysicist is the slow secular change in the mean pole, determined by averaging out both the Chandler and the annual term. Markowitz (1968) finds a motion in a generally similar direction since 1900 of about 0.006"/year. This is of the order of the rate of polar wandering found from palaeomagnetic date ($1/5°$/year). Of course the observed palaeomagnetic pole wandering curve from any one continent (or microplate) results from continental drift relative to other continents as well as from polar wandering. A meaning can be attached to the latter, but in any case both phenomena are caused by slow flow in the solid earth's mantle resulting from solid state creep. Studies of sea floor spreading resolves the motion of the continents on the 10^5 y - 10^7 y time scale, and the palaeomagnetic date from the continents has even less time resolution. Thus the mean motion of the pole as determined by astronomical and other sensitive methods in historic time are an important contribution to the yet poorly understood mechanism of plate motions.

The study of the earth's rotation throws a unique light on the dynamics of the earth's interior and is an important contribution to the geophysicist's programme which in the last quarter of a century has replaced the classical static model of the earth's interior by a dynamical one.

When the motion of the Moon in longitude is expressed in dynamical time, it is found to be a parabolic function of time. There is thus a real acceleration of the Moon. Until recently the results for this acceleration obtained from observatory data over the last 300 years and from ancient and medieval eclipse and other observations were thought to differ. However, revaluation of both methods gives essentially the same value. Thus Morrison and Ward (1975) deduced an acceleration of ($"/cy^2$) $- 26 \pm 2$ (corresponding to a rate of retreat of the Moon from the Earth of about 4.9 cm yr^{-1}) from transits of Mercury since AD 1677. Muller (1976) in his revision of the analysis of the ancient and medieval data by Muller & Stephenson (1975) obtained $- 30.0 \pm 3.0$ (4.4 cm yr^{-1}). Daily, monthly and annual growth increments are seen on skeletons of marine creatures. However, further developments in the use of fossils for counting the ratios between these increments is required if these parameters are to be of use in the evolution of the Earth-Moon system (Rosenberg & Runcorn, 1975).

Tidal friction appears adequate to provide the decelerating torque required by the astronomical observations (Lambeck 1978). However, this torque cannot have remained over the earth's life, and it is still doubtful whether it has remained constant over Phanerozoic times in view of changes in oceanic and continental distribution (Pannella 1975).

REFERENCES

Bauer,L.A.1895.Amer.J.Sci.50,pp.109,189,314.
Brouwer,D.1952.Astron.J.57,pp.125.
Cazenave,A. & Lambeck,K.1976.Geophys.J.Roy.Astr.Soc.46,pp.555.
Guinot,B.1972.In Rotation of the Earth, IAU Symposium No.48,Eds.P.
    Melchior and S.Yumi, D.Reidel Publishing Co. pp.46.
Halley,E.1692.Phil.Trans.Roy.Soc. 17, pp. 563.
Hide,R.1969.Nature, 222, pp. 1055.
Jones,H.Spencer.1939.Mon.Not.Roy.Astr.Soc.99, pp.541.
Kahle,A.B., Ball,R.H. & Cain,I.C.1969.Nature,223,pp.165.
Lambeck,K.1968.Phil.Trans.Roy.Soc.A287,pp.545.
Markowitz,W.1968.In Continental Drift, Secular Motion of the Pole and
    Rotation of the Earth,IAU Symposium No.32, Eds.W.Markowitz and
    B.Guinot, D.Reidel Publishing Co. pp.25.
Mintz,Y. & Munk, W.H.1951.Tellus,3,pp.117.
Morrison,L.V. and Ward,C.G.1975.Mon.Not.Roy.Astr.Soc.173,pp.183.
Muller, P.M.1976.Report SP43-36,Jet Propulsion Laboratory, California.
Muller,P.M. & Stephenson,F.R.1975.In Growth Rhythms and the History of
    the Earth's Rotation, Eds.G.D.Rosenberg and S.K.Runcorn, John
    Wiley and Sons, pp.459.
Pannella,G.1975.In Growth Rhythms and the History of the Earth's Rotation,

Eds.G.D.Rosenberg & S.K.Runcorn, John Wiley & Sons, pp.253.
Roberts,P.H.1972.J.Geomag.and Geoelec.24,pp.231.
Rochester,M.G.1960.Phil.Trans.Roy.Soc.A252, pp.531.
Roden,R.B.1963.Geophys.J.Roy.Astr.Soc.7,pp.361.
Rosenberg, G.D. and Runcorn,S.K.(Eds.)1975.Growth Rhythms and the History of the Earth's Rotation,John Wiley and Sons.
Runcorn,S.K.1955.Trans.Amer.Geophys.Un.36,pp.191.
Runcorn,S.K.1969.Science,163,pp.1227.
Runcorn,S.K.1970.In Earthquake Displacement Fields and the Rotation of the Earth.Eds.L.Manshinha,D.E.Smylie and A.E.Beck.D.Reidel Publishing Co. pp.181.
Sitter,W.de 1927.Bull.Astron.Inst.Neth.4,pp.21.
Vestine,E.H.1953.J.Geophys.Res.58,pp.127.
Vestine,E.H., Laporte,L., Large,I., Cooper,C. and Hendrix,W.1947. Description of the Earth's Main Magnetic Field and its Secular Change, 1905-1945.Carnegie Inst.Wash.Publ.785.

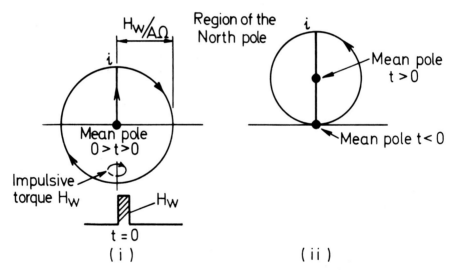

EXCITATION OF CHANDLER WOBBLE

(i) Impulsive torque from core.   (ii) Mass displacement by earthquake.

ESTIMATION OF THE PARAMETERS OF THE EARTH'S POLAR MOTION

Clark R. Wilson
Department of Geological Sciences
The University of Texas at Austin

The parameters of the Earth's polar motion have been estimated by a method which accounts for observational error and which treats two possible statistical models for the polar motion excitation process. Both a 100 year and a 78 year long polar motion series have been used to estimate the Chandler wobble frequency, F_c, and quality factor, Q_c; the noise level in the polar motion data; and the magnitude of the excitation process. Discussion of the annual component of polar motion has been omitted because its frequency (1 cycle per year (cpy)) and cause (annual motion of air and water) are presumed known.

1. INTRODUCTION

A traditional approach to estimating the polar motion parameters, particularly F_c and Q_c, has been to use the periodogram method of fitting sines and cosines of various frequencies to the polar motion data. However, in 1940 Sir Harold Jeffreys suggested that since the wobble excitation process is likely to be random, a statistical estimation method ought to be superior to periodogram analysis, because a randomly excited damped oscillator will not display purely sinusoidal behavior. The estimation method used here is based upon such a statistical approach.

If the polar motion excitation behaved as an independent random process, then its Fourier power spectrum would be white, containing the same variance at all frequencies. One would then expect the polar motion power spectrum to show a symmetric peak at the Chandler frequency, like the dotted line in Figure 1. The solid line in Figure 1 shows that the spectrum of the polar motion data (for the years 1901-1970, with annual term removed) is similar to the dotted curve near the Chandler frequency, but much larger elsewhere. Apart from the Chandler frequency peak, the spectrum of the data is best described as "red" since it rises toward zero frequency. It appears that the data is contaminated by a substantial amount of "red" noise. Rather than introducing separate corrections to account for the presence of this noise, I will estimate noise and polar motion parameters simultaneously, thereby hopefully eliminating any noise bias.

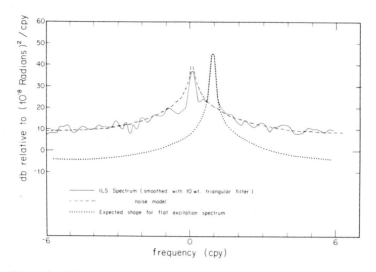

Fig. 1. The power spectrum of the 1901-1970 ILS series suggests that a substantial amount of red noise is contaminating the data.

2. THE DATA

The polar motion time series used in this study has been compiled from the following sources for the periods indicated: 1878-1977 data were taken from the tabulation by Rykhlova (1969). A gap of several years duration in the component along 90° east longitude was interpolated by hand using the Greenwich component as a guide. 1890-1899 data were taken from Stoyko (1972, Tables 7a, 7b). 1900-1967 data were the ILS pole positions reported by Vicente and Yumi (1969, 1970). 1968-1977 data were the 5-day raw pole positions given in the BIH annual reports and Circular D. The 5-day data were smoothed using a 7 weight triangular filter in preparation for the interpolation described below.

Adjustments were made to the means of the 1878-1899 and 1890-1899 series to give a smooth transition where they were joined with each other and with the ILS series. A cubic spline interpolator was applied separately to the two components of the 1878-1977 series to resample them at uniform intervals of 1/12 year, resulting in approximately mid-month samples. Annual and higher harmonic components were removed by subtracting the means of the same-named months from the monthly samples.

3. ESTIMATION OF THE PARAMETERS FROM THE DATA

3.1. Notation and Fundamental Equations

Let M_t be the complex valued time series of true polar motion (uncontaminated by noise), and let X_t be the complex valued time series of the position of the excitation axis. Variation of the real part of M_t or X_t represents motion along the Greenwich meridian, while variation of the complex part represents motion along the 90° East meridian. The relationship between M_t and X_t is

$$M_t = -\alpha X_t + e^{\alpha} M_{t-1}, \tag{1}$$

where $\alpha = 2\pi i F_c T(1+i/2Q_c)$, $i = \sqrt{-1}$, and T is the sample interval. $Z = M_t + N_t$ are the data which are contaminated by the noise N_t. I will assume that N_t satisfies the equation

$$N_t = N\eta_t + N_{t-1}, \tag{2}$$

where N is a real-valued constant and η_t is a unit variance, complex-valued white noise process whose real and imaginary parts are independent, identically distributed, zero mean, Gaussian random variables. The dashed line in Figure 1 shows the spectral shape of a time series obeying (2) and demonstrates the similarity with the ILS spectrum apart from the Chandler frequency peak.

Letting the first difference of the data be $D_t = Z_t - Z_{t-1}$ and using (2) to describe N_t, D_t satisfies the equation

$$D_t = -\alpha X_t + \alpha X_{t-1} + N\eta_t - e^{\alpha} N\eta_{t-1} + e^{\alpha} D_{t-1}. \tag{3}$$

3.2. Two Models for the Excitation Process

Model 1: The simplest assumption that might be made about X_t is that it is described by the equation

$$X_t = X\varepsilon_t, \quad \text{(Model 1)} \tag{4}$$

where ε_t is a unit variance white noise process and X is a real-valued constant. Wilson and Haubrich (1976) show that Model 1 is an appropriate assumption if atmospheric variation is the main contributor to X_t.

Model 2: If earthquakes are a major contributor to X_t, as suggested by O'Connell and Dziewonski (1976), then the power spectrum of X_t is probably red. A simple model for X_t in this event is that

$$X_t = X\varepsilon_t + X_{t-1}, \quad \text{(Model 2)} \tag{5}$$

where X is a real-valued constant (but with a value different from that of Model 1) and ε_t is again a unit variance white noise process.

3.3. Estimating Parameters and Confidence Intervals

Box and Jenkins (1970, p. 122) show that (3) has a statistically equivalent description as

$$U_t = D_t - e^{\alpha}D_{t-1} - (A/V)U_{t-1}, \tag{6}$$

where A and V are chosen to make the autocovariances of D_t the same in both (3) and (5), and where U_t is a white noise process with variance V.

For Model 1: $A = -(|\alpha|^2 X^2 + e^{\alpha}N)$;

$V = .5[h_1 + (h_1^2 - 4A^2)^{\frac{1}{2}}]$, where $h_1 = (2|\alpha|^2 X^2 + N^2 + |e^{\alpha}|^2 N^2)$.

For Model 2: $A = -(e^{\alpha}N^2)$;

$V = .5[h_2 + (h_2^2 - 4A^2)^{\frac{1}{2}}]$, where $h_2 = (|\alpha|^2 X^2 + N^2 + |e^{\alpha}|^2 N^2)$.

For given values of the polar motion parameters, U_t is generated recursively from the data using equation (6) with a starting value U_0 of zero. By searching with a computer, one may find parameter values which minimize the variance of U_t and hence are the least squares and maximum likelihood estimates. The search is not difficult because Q_c and F_c are already fairly well known, and because there is only one additional parameter (X/N) contained in (A/V) in (6). X and N are determined separately from the final minimum variance of U_t. The minimum variance of U_t measures how well the assumed model fits the data, with a smaller value indicating a better fit.

Monte Carlo experiments with artificial data were used to demonstrate that Model 2 estimates of all parameters are unbiased. Estimates of Q_c, F_c, and N made using Model 1 were also unbiased, but Model 1 consistently estimated X to be one quarter of its true value. The values in Table 1 have been corrected for this bias. Intervals of confidence were obtained using Box and Jenkins' method (1970, p. 229) which examines changes in the variance of U_t as a function of Q_c, F_c, and X/N.

4. RESULTS AND CONCLUSIONS

Table 1 shows estimates of the various parameters obtained from the entire 1878-1977 series and from the 1900-1977 portion, and also gives power spectral densities of X_t at the frequency, F_c, denoted by $S_x(F_c)$. Estimates of Q_c, F_c, N, and $S_x(F_c)$ obtained from both time series and both models are approximately the same.

Model 1 was found to fit both time series slightly better than Model 2, suggesting that X_t has a white rather than a red power spectrum. Both

Table 1. Estimates and Intervals of 90% Confidence.

Data	X 10^{-8} Radians	$S_x(F_c)$ (10^{-8} Radians^2/cpy)	Q_c Dimensionless	F_c cpy	N 10^{-8} Radians
		Model 1 - White Excitation Spectrum			
1879-1977	15.3 (13.1-18.5)	19.5 (14.3-28.6)	59 (26->1000)	.848 (.841-.857)	19.6 (19.3-19.9)
1900-1977	13.6 (11.2-16.4)	15.5 (10.5-22.3)	61 (27->1000)	.845 (.837-.853)	18.4 (18.2-18.7)
Meteorological Variation (Wilson and Haubrich, 1976)	7-10	5-8			
		Model 2 - Red Excitation Spectrum			
1878-1977	6.9 (5.9-8.2)	20.7 (15.1-29.2)	60 (27->1000)	.849 (.841-.857)	19.7 (19.4-20.0)
1900-1977	6.1 (5.2-7.2)	16.3 (11.6-22.7)	62 (27->1000)	.845 (.837-.853)	18.5 (18.3-18.8)
Earthquakes (O'Connell and Dziewonski, 1976)	2	2			

Models 1 and 2 fit the 1900-1977 series considerably better than the 1878-1977 series, suggesting that the pre-1900 data do not have the same statistical description as the later data.

Estimates of F_c shown in Table 1 are slightly larger than the one obtained by Jeffreys (1972), perhaps because Jeffreys did not use precisely the same data set. The difference may also be due to the fact that red noise in the data would tend to make Jeffreys' estimates of F_c too small, while the estimates in Table 1 are presumably not biased in this way.

Estimates of Q_c are nearly the same as Jeffreys' (1972) value. As shown in Table 1, the upper confidence limit for Q_c exceeds 1000 for both models and both time series. However, Q_c is probably not much larger than 1000, due to dissipation in the mantle and oceans. Thus, it appears that physical reasoning may provide a more stringent upper bound on Q_c than does the polar motion data. For Model 1, Q is <u>less</u> well determined for the longer time series, since the confidence interval is slightly wider. This suggests that the pre-1900 data has not been used properly and that it needs to be analyzed in some other way.

From the Model 2 results, the earthquake effect is too small by a factor of about 3 in amplitude (X) or a factor of 9 in power $(S_x(F_c))$ to account for the Chandler wobble excitation. From the Model 1 results, the estimated meteorological effect is too small by a factor of roughly 2 or 3 in power, or $\sqrt{2}$ to $\sqrt{3}$ in amplitude to account for the Chandler wobble excitation. Since there is some evidence for correlation between atmospheric variation and polar motion, as shown by Wilson and Haubrich (1976), the discrepancy in amplitude is perhaps the best measure of how much of the Chandler wobble excitation remains to be explained.

REFERENCES

Box, G. and Jenkins, G.: 1970, "Time Series Analysis, Forecasting and Control", Holden Day, San Francisco.
Jeffreys, H.: 1940, Mon. Notices Roy. Astron. Soc. 100, p. 139.
Jeffreys, H.: 1972, in S. Yumi and P. Melchior (eds.), "The Rotation of the Earth", D. Reidel, Dordrecht.
O'Connell, R. and Dziewonski, A.: 1976, Nature 262, pp. 259-262.
Rykhlova, L.: 1969, Soviet Astronomy 12, p. 898.
Stoyko, A.: 1972, in A. Beer (ed.), "Vistas in Astronomy", vol. 13, p. 51.
Vicente, R. and Yumi, S.: 1969, Publ. Int. Latitude Obs. Mizusawa 7, pp. 41-50.
Vicente, R. and Yumi, S.: 1970, Publ. Int. Latitude Obs. Mizusawa 7, p. 109.
Wilson, C. and Haubrich, R.: 1976, Geophys. J. Roy. Astron. Soc. 46, pp. 707-743.

ROTATIONAL VELOCITY OF AN EARTH MODEL WITH A LIQUID CORE

S. Takagi
International Latitude Observatory
Mizusawa, Iwate
023 Japan

ABSTRACT

There have been many papers discussing the rotation of the Earth (Jeffreys and Vicente, 1957; Molodenskij, 1961; Rochester, 1973; Smith, 1974; Shen and Mansinha, 1976). This report summarizes the application of the perturbation method of celestial mechanics to calculate the rotation of the Earth (Takagi, 1978). In this solution the Earth is assumed to consist of three components: a mantle, liquid outer core, and a solid inner core, each having a separate rotational velocity vector. Hamiltonian equations of motion were constructed to solve the rotational motion of the Earth.

Using the results of Jacobs (1976) and that of Shen and Mansinha (1976), the perturbation of the rotational velocity can be expressed as the result of superficial forces acting over the boundary surfaces, terms due to dynamical processes in the Earth, terms due to the differences among the amplitudes of the motions of the rotation axes of the three components, terms due to the non-coincidence of the rotation axes, and terms due to the internal motion in the outer core. The results may be summarized as follows:
a. The dissipative terms at the boundaries have been discussed in detail by Crossley and Smylie (1975).
b. The term due to body forces shows periodic motion, but it disappears for diurnal nutation.
c. The effects of deformation show periodic variation, but only a term due to the outer core effects diurnal nutation.
d. A periodic term arises from the difference between the amplitudes of the motion of the rotation axes of the three components.
e. The discrepancy among the directions of the rotation axes causes secular terms.
f. Motion in the outer core causes periodic motion, but this disappears for diurnal nutation.

It should be noted that periodic motion in the rotational velocity is expressed in one of two forms: either $\int \sin\theta \sin(\sigma t + \psi) dt$ or $\int \sin(\sigma t + \psi) dt$, where σ is the frequency of the perturbing effective torque, θ is the precession and nutation in obliquity and ψ that in longitude.

REFERENCES

Crossley, D. J. and Smylie, D. E.: 1975, Geophys. J. Roy. Astron. Soc. 42, pp. 1011-1033.
Jacobs, J. A.: 1976, "The Earth's Core", Cambridge University Press, London.
Jeffreys, H. and Vicente, R. O.: 1957, Mon. Not. Roy. Astron. Soc. 117, pp. 142-161 and pp. 162-173.
Molodenskij, M. S.: 1961, Commun. Obs. Royal Belgique No. 188.
Rochester, M. G.: 1973, EOS Trans. American Geophys. Union 54, pp. 769-780.
Shen, Po-Yu and Mansinha, L.: 1976, Geophys. J. Roy. Astron. Soc. 46, pp. 467-496.
Smith, M. L.: 1974, Geophys. J. Roy. Astron. Soc. 37, pp. 491-526.
Takagi, S.: 1978, Publ. Int. Latitude Obs. Mizusawa 12, in press.

ON INVESTIGATIONS OF THE TIDAL WAVES M_f AND M_m

G. P. Pil'nik
Sternberg Astronomical Institute
Moscow, U.S.S.R.

ABSTRACT

The comparison of astronomical time observations with the theory of solid-Earth tides makes it possible to determine the Love number, k, which characterizes the elastic properties of the Earth. In addition, the comparison of values of k determined from different tidal waves allows us to judge the accuracy of the nutational theory in astronomical observations since both tides and the Earth's nutation are produced by the same causes.

The non-uniformity of the Earth's rotation has been investigated previously (Djurovic, 1975, 1976; Guinot, 1970, 1974; Pil'nik, 1970, 1974, 1975, 1976) to derive k. These works show that the Love number derived from the M_f tide, $k(M_f)$, is greater than that for the M_m tide, $k(M_m)$. In this investigation 9103 residuals of observed Universal Time made from 1951 to 1975 were analyzed. This extended series of observations was based on Standard Time results from 1951 to 1967 and Bureau International de l'Heure results from 1968 to 1975. Only the M_f and M_m tidal waves were investigated using Tukey's method of spectral analysis. The Love numbers were estimated using the methods of Pil'nik (1974).

This analysis yields the estimates for the Love numbers, $k(M_f) = 0.301 \pm 0.005$, and $k(M_m) = 0.282 \pm 0.004$. A phase shift in the M_f wave is found to change considerably, depending on the length of the data record analyzed. The elliptical wave M_m is delayed in phase by more than one day.

REFERENCES

Guinot, B.: 1970, Astron. Astrophys. 8, p. 1.
Guinot, B.: 1974, Astron. Astrophys. 36, p. 1.
Djurovic, D.: 1975, Obs. Roy. Belgique Bull. Observations Marees
    Terrestres, vol. 4, pp. 3-5.
Djurovic, D.: 1976, Astron. Astrophys. 47, p. 3.

Pil'nik, G. P.: 1970, Astron. Zh. Akad. Nauk SSSR 47, p. 1308.
Pil'nik, G. P.: 1974, Izvestia Akad. Nauk SSSR, Fizika Zemli, No. 4, p. 3.
Pil'nik, G. P.: 1975, Astron. Zh. Akad. Nauk SSSR 52, p. 178.
Pil'nik, G. P.: 1976, Astron. Zh. Akad. Nauk SSSR 53, p. 889.

DISCUSSION

B. Guinot: In similar studies, using the same method as Pil'nik, I found a strong dependence of the value of k derived from the M_f wave on the position in time of the interval under consideration. Has Pil'nik found such a phenomenon?

Ya. S. Yatskiv: Pil'nik uses a technique of direct comparsion between theoretical results and observations; it is not necessary to be able to separate the contributions of the M_f and M_m waves.

N. P. J. O'Hora: Over how long an interval does the series of observations extend? A 20-year series is long enough to enable the terms at 13.66 and 13.63 days to be resolved from each other.

Ya. S. Yatskiv: That is true if you are using the method of least squares, but in the method employed here the window prevents separation of these terms.

OCEANIC TIDAL FRICTION: PRINCIPLES AND NEW RESULTS

P. Brosche
Sternwarte der
Universität Bonn

J. Sündermann
Lehrstuhl fur Strömungsmechanik
der TU Hannover

Federal Republic of Germany

The main problems of the hydrodynamical integrations are analyzed. New results are presented for the effect of an extreme ice age.

1. PRINCIPLES

It should be recognized that the angular momentum transfer between Earth and Moon via the hydrosphere has essentially two constituents:
a) The interaction between the Moon and water of the oceans, mediated by the bodily tidal forces. In order to compute the time-averaged torque around the axis of rotation of the Earth, we need only the east-west tangential component, F, of the space density of the force; since F is purely periodical, its time average vanishes. Consequently, we need the tidal elevation part, ζ, of the water depth alone. Then the (time-averaged) net torque, L, on the water turns out to be

$$L_1 = \int_{\text{surface of the oceans}} \int_{\text{tidal period}} \rho[(R \cos \phi) \zeta F] dt\, dq, \qquad (1)$$

(R = radius of the Earth, ϕ = latitude, ρ = density of the water, dq and dt are the surface and time differentials). Note that the essential part is the product, ζF.

b) The interaction between the water and the solid Earth mediated by the surface forces of bottom friction. The most realistic representation for the absolute value of the surface density of these forces is thought to be given by the empirical law

$$K^* = r\rho w^2, \qquad (2)$$

where w is the velocity of the water relative to the ground, and r a dimensionless constant (in our computations r = 0.003). With u and v being the east-west and north-south components of w, we can write the east-west component of K*

$$K = K^* u/w = \rho u w = \rho u (u^2 + v^2)^{1/2}. \qquad (3)$$

In this case the net torque acting on the water (and, with reversed sign, on the solid Earth) is

$$L_2 = \int_{\substack{\text{surface of} \\ \text{the ocean} \\ \text{bottom}}} \int_{\substack{\text{tidal} \\ \text{period}}} [(R \cos \phi) K] dt \, dq. \qquad (4)$$

The essential "hard core" here is the product uw.

Since the water is neither a sink nor a source of angular momentum, we have

$$L_1 + L_2 = 0. \qquad (5)$$

Consequently, the determination of either L_1 or L_2 suffices computationally, but for a complete understanding (and for a test of the computations) the independent determination of both L_1 and L_2 is very desirable. This would seem conceivable using oceanographic observations, but for the time being these are too sparse. At present, we have to rely on theoretical models of oceanic tides. It seems doubtful whether such models can provide independent estimates of L_1 and L_2 because, in general, we expect that only those which include both kinds of interactions will give correct results. This "all or nothing" point of view is, however - and luckily! - not supported by the computational experience. Models without any bottom friction or those which use Laplace's tidal equation and hence use precise harmonic motions lead to $L_2 = 0$ but an astonishingly realistic $L_1 \neq 0$. Since these models violate equation (5) they cannot be stationary in a strict sense. Our models contain the friction according to equation (2) and are therefore, in principle, able to represent both kinds of interactions. In practice, L_1 quickly converged to a stationary value after a few tidal cycles of iterations in which the tidal movements themselves reached a degree of stationarity which is sufficient for usual oceanographic purposes. On the contrary, L_2 needed many more iterations and an unprecedented accuracy to reach even the right order of magnitude.

We can still not give a complete explanation, but two heuristic points are undoubtedly near to the cause: First, the ratio between the terms of the bottom friction and of the pressure gradient in the hydrodynamical equations depends strongly on the depth in such a way that the friction is comparable with the latter only for depths up to several tens of meters (Sündermann and Brosche, 1978). Therefore the bottom friction can manifest itself sufficiently quickly only in the case of shallow seas. Second, the time behavior of the tidal elevation, ζ, and of the tidal velocities, u and v, is mainly harmonic. That is, within the coefficients of a Fourier expansion, the two coefficients belonging to the tidal frequency dominate over all others. Because of the orthogonality of Fourier series, we have Parseval's equation for the time average of a product xy:

$$\overline{xy} = X_0 Y_0 + \tfrac{1}{2} \sum_{\nu=1}^{\infty} (X_\nu Y_\nu + X'_\nu Y'_\nu), \tag{6}$$

where X_0, Y_0 are the constants and X_ν, X'_ν, Y_ν, Y'_ν, the coefficients of the sine and cosine terms. Looking into equation (1), we recognize that the time average, $\overline{\zeta F}$, can be reduced to $X_1 Y_1 + X'_1 Y'_1$ because the force is, by definition, precisely harmonic. Since only the main term in ζ is involved, we get the desired result from first order quantities. In contrast, the Fourier expansion of w contains only even terms ($\nu = 0$, 2, 4, ...) if u and v are purely harmonic (w being, then, the radius in the elliptic path of the velocity vector). The presence of second order terms other than $\nu = 1$ in u and v leads to the occurrence of odd terms of second order in w (and also to small changes of the even terms). Then the time average of uw in equations (3) and (4) becomes a series where all the terms contain a second order factor alternately arising from u or w! In summing up, the numerical solution of the hydrodynamical equations has to be correct in the first order quantities to get a meaningful result from equation (1), but the second order accuracy is necessary for obtaining such a result from equation (4). The numerical results presented in the following section for global ocean models are therefore based on applications of equation (1).

2. NEW RESULTS

So far, all our results refer to the M_2 tide and to a schematic allowance for the elasticity of the Earth by using a reduction factor 0.69 for the effective tidal constant. The resulting torque for our model of the present oceans is in good accordance with astronomical values. As a contrast, we have treated models of the Pangea situation of the continents. We obtained smaller values than for the present configuration, which is desirable in order to prevent an apocalyptic Gerstenkorn event (Brosche and Sündermann, 1977; Sündermann and Brosche, 1978).

A plausible reason for considerable variations of the torque acting in the more recent geological past is the alternation between ice ages and intervening periods. In the case of an ice age more sea water is tied up in the polar ice caps than today and the sea level is consequently lower. While the relative topography is practically the same as at present, it is sufficient as a first approximation to lower the sea level in this case by 100 m for the purpose of representing an extreme ice age. The corresponding tides are very similar to those of the present situation; the latitude distribution is somewhat less concentrated towards an equatorial belt than that of the present tides. The main result is that the average torque is only a few percent greater than the present value:

$$L_1 = -5.2 \times 10^{23} \text{ dyn cm}.$$

Thus we can conclude that at least for more recent geological epochs the variation of the glaciation of the Earth does not seem to complicate the reconstruction of the Earth's rotation. This need not be true for every epoch, because, as we learned from the Permian models, the opening or blocking of crucial passages is of importance for our aims.

REFERENCES

Brosche, P. and Sündermann, J.: 1977, in J. D. Mulholland (ed.), "Scientific Applications of Lunar Laser Ranging", D. Reidel Publ. Co., Dordrecht, Boston, pp. 133-141.
Sündermann, J. and Brosche, P.: 1978, in P. Brosche and J. Sündermann (eds.), "Tidal Friction and the Earth's Rotation", Springer, Berlin, Heidelberg, New York, (in press).

THE STABILITY OF CONTINENTAL BLOCKS IN SEISMICALLY ACTIVE REGIONS

V. P. Shcheglov
Astronomical Observatory of Tashkent
U.S.S.R.

ABSTRACT

In connection with the problem of continental drift some authors voice a supposition of the existence of progressive and rotational motion of continental blocks, particularly in seismically active regions. Azimuths of some terrestrial objects determined over long time intervals were used to investigate these motions. Naturally, the reliability of the result depends on the length of these intervals.

The axis of the meridian instrument of Uloug-Beg Observatory in Samarkand, installed 550 years ago (1420-30) is the most ancient meridian direction on the Earth. Using measurements made by the author in 1941, 1956, and present-day observations, it was found that the direction of the axis had moved through an angle of 7'.5. Allowing for observational errors in the pre-optical period one can accept a change of 5' or 0".5 per year. To ascertain if the observed change is caused by rotation of the continental block or local deformation repeated measurements of azimuths of triangulation stations first measured forty years ago were made. The results show the absence of continental block rotation in the region of Middle Asia within a possible accidental error of ±0".5.

DISCUSSION

S. K. Runcorn: Rotation of large plates associated with continental drift is of the order of 10^{-7} per year, but local rotation of small blocks associated with seismic areas may be of the order found from the ancient discrepancy of the meridian. May it not be that the region is just quiescent at present? Such motions are likely to be discontinuous but may add up to considerable values over 500 years.

V. P. Shcheglov: This is a seismic zone. The effect is not a local rotation.

General discussion: Session 1

Chairman: Dr G A Wilkins

The chairman invited the Symposium to consider whether the IAU ought to take action to improve the effectiveness of work on time and the rotation of the Earth; was there a need for formal adoption of standard reference systems, or for changes in the time systems in use? There appeared to be a need to improve arrangements for coordination in the use of new techniques, and consideration should be given to the procedures for the reduction, analysis and publication of the results obtained. A draft resolution concerning possible amalgamation of Commission 19 and 31 had been submitted by J.D. Mulholland; it would be displayed, and discussed at the final session.

The chairman then called B. Guinot, who summarized arguments, presented in his invited paper, in favour of the adoption of a new system of reference coordinates in which universal time would be directly proportional to "stellar angle".

During the following discussions it was emphasized that the implications of such a change for different classes of user would need careful study; if firm proposals could be produced before the General Assembly then decisions might be taken there, but it would not be appropriate to hold detailed discussions immediately. Adopted concepts ought to be readily comprehensible to workers in other fields, and reference points should be observable, at least in principle.

Time scales for the description of Earth rotation were discussed next. B. Guinot noted that the adoption of FK5 would introduce a step in UT1, and suggested that it might be desirable to introduce a variant of universal time (UT3 ?) from which the effects of zonal tides had been removed. This would reduce problems of interpolation from the 5-day means; at present some short-period terms had amplitudes comparable with the residual scatter. Publication of UT2, on the other hand, appeared to be unnecessary; the conventional terms used in its formation do not represent well the variable effects that are observed, and there is a danger that users will employ UT2 when they need UT1.

In the discussion it was recognised that although UT2 is useful for work within the time services, this does not require formal publication of tabulated values of UT2.

Views expressed during discussions on internation coordination were:-

- New facilities ought not to be planned in isolation, but there are at present no formal arrangments for international coordination.

- The IAU could and should influence the development of new systems; it should not, as too often in the past, restrict its efforts to the integration of systems already in operation.

- The new techniques operate on a global scale, some requiring observations from only a few stations working as a single system. It is not obvious that IAU involvement would be helpful within such a system, but the IAU might usefully aim to coordinate periods of intense effort by several independent systems using various techniques.

- International campaigns of this kind could perhaps add weight to applications for observing time on some equipment for which Earth rotation studies would normally have low priority, but the campaigns might seem merely irrelevant or irritating to operators of other systems already dedicated to time-critical observations, sometimes scheduled years in advance.

- It is difficult to use observations which are made only intermittently or are interrupted. There is a need for continuity, and for concurrent daily operation of independent systems over periods whose durations should not be limited in advance, but should be determined only after intercomparison of the results obtained.

- Analysis of observations ought not to be undertaken by isolated groups using independently developed methods, but there could be dangers in too much formal organisation of effort; it is important that observations should be processed and the results published quickly. Comparison should normally be with the results of the BIH or IPMS, and the results of retrospective analyses should also be made available to these bodies.

General discussion: Session II

Chairman: Dr G A Wilkins

The Symposium discussed whether the same reference frames could adequately serve the needs of astronomers, geodesists and geophysicists. The following points were made:-

- It is desirable to 'freeze" the system; that is, to adopt reference models which will not be subject to further change. Adoption of a rigid-body model would admittedly cause difficulties, but some speakers felt that, with the compensating advantage of simplicity, these would be preferable to the succession of changes which would be likely to result from the use of deformable models tailored to current knowledge.

- Although the Earth-rotation models were linked to directions rather than points, they would be fit for use in geodesy if supplemented by distances and an origin.

- It seemed possible that it might be desirable to annul the resolutions adopted the previous year by Symposium 78 at Kiev, although others felt that the second might stand since the system adopted was not strongly model dependent. A working group had been set up to make recommendations on the treatment of nutation and it was not unlikely that these might be incompatible with the Kiev resolutions. Views on the matter should be sent to the Chairman of the group, P.K. Seidelmann at USNO.

General discussion: Session III

Joint chairmen: R.O. Vicente, Dr G.A. Wilkins

R.O. Vicente drew attention to the completion by the International Latitude Observatory of a retrospective analysis of International Latitude Service data, and proposed Resolution 1. The resolution was adopted.

The Symposium also adopted Resolution 2, calling for the appointment of a working group to evaluate new techniques and make recommendations for a new international programme. The hope was expressed that information would be exchanged between this and other working groups, for example that of COSPAR on lunar laser ranging.

No further action was taken concerning the adopted position of the pole; it was stated that the almanac offices would need nutation terms by 1980 in order that these could be incorporated in the almanac for 1984, and that ideas or comments should be made known to P.K. Seidelmann at USNO.

The final subject discussed was the desirability of merging Commissions 19 and 31, which had jointly sponsored the Symposium. The discussion was held at the request of J.D. Mulholland, seconded by K. Johnston. They stated that several members of the IAU having astronomical interests both within and outside the fields of these two commissions were unable to be members of both because of the overall restriction to membership of not more than 3 commissions; information was consequently not reaching all to whom it was of interest. There was already a large common membership and many meetings were arranged jointly. It was believed that mergers between other commissions were being considered for similar reasons.

Other speakers stated that the possibility of a merger had been considered previously and rejected, although it might justifiably be held that circumstances had changed; in particular, atomic time was now widely accessible and coordination procedures had been established. The problems considered by the two commissions were, however, quite different and the quality of technical decisions would not be improved by merging. There were, moreover, other interests, notably those of the standards laboratories, which ought to be represented in Commission 31; it would be more difficult to accommodate them in a merged commission.

After further discussion a motion requesting both Commissions to conduct polls of their members on the subject of a possible merger, and to make the results of the polls known at the General Assembly in Montreal, was rejected by 25 votes to 16.

RESOLUTIONS

Resolution No. 1

IAU Symposium No. 82 expresses its satisfaction with the methods employed by the International Latitude Observatory of Mizusawa in compiling and re-reducing the past International Latitude Service data.

The symposium further expresses its gratitude to the staff of the International Latitude Observatory of Mizusawa in recognition of the magnitude of the task.

Resolution No. 2

IAU Symposium No. 82 recommends that the presidents of Commissions 19 and 31 appoint a working group to promote a comparative evaluation of the techniques for the determination of the rotation of the Earth and to make recommendations for a new international programme for observation and analysis in order to provide high-quality data for practical applications and fundamental geophysical studies.

WORKING GROUP MEMBERSHIP

Chairman: G. A. Wilkins (UK)
Lunar Laser: E. C. Silverberg (USA)
           Y. Kozai (Japan)
           Yu. L. Kokurin (USSR)
Satellite Laser: D. E. Smith (USA)
           G. Veis (Greece)
           L. Aardoom (Holland)
Doppler: R. J. Anderle (USA)
       F. Nouel (France)
Radio Interferometry: W. C. Melbourne (USA)
           W. E. Carter (USA)
           B. Elsmore (UK)
Classical Methods: Y. S. Yatskiv (USSR)
           R. O. Vicente (Portugal)
BIH Representative: M. Feissel (France)
IPMS Representative: K. Yokoyama (Japan)

INDEX

aberration 15, 48, 112
adjustment 139ff.
astrolabe 41ff., 85ff., 129ff., 135, 139, 143, see also Danjon astrolabe
astrometric satellite 157
atmosphere 178, 185
atmospheric drag 232, 240ff., 265
atmospheric effects on the Earth's rotation 61ff., 110, 302, 309
axis of angular momentum 8, 92, 169ff.
axis of figure 8, 93, 115ff., 166ff.
axis of inertia 115ff.
axis of rotation 8, 92, 169, 313, 317

Beacon Explorer C 239
body tide 67ff.
Bureau International de l'Heure (BIH) 2, 6, 23, 29, 42, 47, 76, 83, 109, 125, 138, 145, 138, 145, 213, 231, 235, 239, 257, 268, 279, 298, 308, 315, 324
Bureau International de l'Heure (BIH) 1968 system 29, 47, 96, 142, 175, 201

cesium beam frequency standards 19
Chandler free motion of the pole 2, 120, 145, 263, 269, 281, 301ff., 307
continental drift 110, 253, 321
Conventional International Origin (CIO) 29, 48, 95ff., 114, 123, 175, 183, 240, 265
core-mantle coupling 55, 301ff.
core of the Earth 201, 313
correlation functions 5
covariance analysis 227ff.

Danjon astrolabe 47, 76, 110, 137, see also astrolabe
data analysis 129ff., 139ff., 290ff., 297
data transmission 293
deep space network 199ff., 226ff., 293
diurnal nutation 15, 35, 79, 112, 169ff., 313
Doppler derived polar coordinates 74, 75, 110, 212, 263ff., 279ff., 287ff., 295ff.
Doppler Polar Motion Service (DPMS) 145, 212, 298
dynamically defined systems 152

earthquakes 191, 243, 246, 309
Earth's gravity field 231, 240, 265
Earth models 3, 44, 167, 169ff., 301ff., 313

Earth tides 15, 25, 36, 50, 79, 110, 187, 200, 231, 240, 245, 261, 265, 315
ephemeris Earth 167
ephemeris reference frame 165ff
equinox 9
EROLD campaign 251, 257ff., 261
excitation function 279, 281, 307ff.

filters 37
fundamental astrometry 165ff.
fundamental star catalog 9, 85ff., 91, 110, 155ff., 201

general relativity 152, 201
geomagnetic effects 301ff.
GEOS-3 239ff.

hydrogen frequency standards 19, 20, 178, 194, 206

intermediary systems 161, 168
international atomic time, see Temps Atomique International (TAI)
international coordination 324
International Latitude Service (ILS) 2, 83, 103, 127, 145, 268, 308, 326
International Polar Motion Service (IPMS) 76, 83, 125, 145, 257, 268, 324
ionosphere 179, 185, 202

Jeffreys-Atkinson axis 92, 168

kinematically defined systems 153

LAGEOS 155, 231ff., 243, 245, 260
laser ranging 75, 99, 175
laser ranging to satellites 155, 231ff., 239ff., 245, 247
length of day 53
Liouville equation 116
local effects 110, 132, 158, 262, 321
longitude reference points 139ff.
Loran-C 19
Love numbers 50, 67ff., 79, 170, 187, 265 315
lunar laser ranging 110, 154, 245, 247ff., 257ff., 261
lunar librations 154, 253
lunar occultations 55, 155

331

lunar orbit 81, 249, 253
lunisolar torque 170

mean latitude 29, 123, 125
mean pole of epoch 29, 97
MEDOC campaign 287ff., 295ff.

nutation 2, 8, 35, 41ff., 79ff., 91, 104ff., 110, 132ff., 148, 169ff., 188, 200, 313, 315, 325, 326, see also diurnal nutation

ocean tides 67ff., 231, 317
orbit determination 231ff., 239, 264ff., 291, 297

photographic zenith tube 42, 47, 65, 67ff., 75, 79, 110
plate motion 15, 159, 191, 253, 304
polar coordinates 37, 75ff., 103ff., 125, 145, 234ff., 241, 263ff., 279
polar motion 29ff., 79, 89, 145, 179, 191, 199, 231ff., 239ff., 245, 247, 253, 262, 263ff., 279ff., 295ff., 301ff., 307ff.
Polaris project 191ff.
precession 1, 8, 91, 110, 170, 188, 200, 313
proper motions 15, 104, 110

quality factor 307

radiation pressure 231, 265
radio interferometry 74, 75, 98, 110, 156, 175, 177ff., 183ff., 191ff., 199ff., 211ff., 225ff., 245
radio sources 156, 163, 179, 187, 204, 212, 225ff.
radiometer 186
reference systems 7ff., 29ff., 32ff., 89ff., 103, 109, 151ff., 165ff., 175, 183, 192, 201, 204, 231, 247, 265, 280, 298, 323

refraction 65, 104, 113, 129ff., 240, 245
relativistic deflection of electromagnetic waves 15, 188
rotational velocity 53, 59, 313

secular acceleration of the Moon 153, 155
secular geomagnetic changes 301ff
secular polar motion 29ff., 120, 123, 285, 301ff.
secular variation of latitude 123, 125
seismic regions 321
sidereal time 184
SI system 19
smoothing 24, 68, 76, 130, 132, 137
solar activity 59
special relativity 200
spectral analysis 25, 32, 53, 56, 59, 145ff., 275, 307ff., 315
star catalogs 65, 74, 85ff., 104, 106, 110, 134, 137, see also fundamental star catalog
sway 16

tectonic motions 110, see also plate motion, continental drift
Temps Atomique International (TAI) 12
tidal effects 67ff, 154, 170, 305, 323
tidal friction 317
time scales 19ff., 23ff., 29ff., 47ff.
troposphere 178, 185, 202, 240, 245

Universal Time 12ff., 23ff., 29ff., 47, 59, 73ff., 75ff., 83, 109, 138, 179, 184, 191, 199, 213, 236, 245, 249, 259, 261, 266, 315, 323

variations in the Earth's rotation 2, 4, 25, 49ff., 53, 55, 59, 61ff., 71, 247, 301ff., 317
variation of latitude 1, 2, 125, 138, 145ff.

zenith telescope 76, 104, 110, 123, 135